MARGIN OF ERROR

Matthew Abess

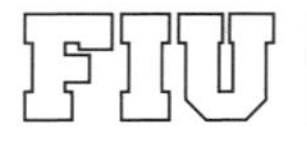

The Wolfsonian
Florida International University

Kara Pickman, editor
Marlene Tosca, art director
Brittany Ballinger, graphic designer
David Almeida and Lynton Gardiner, photographers

The Wolfsonian–Florida International University
1001 Washington Avenue
Miami Beach, FL 33139
wolfsonian.org

The Wolfsonian receives ongoing support from the John S. and James L. Knight Foundation; State of Florida, Department of State, Division of Cultural Affairs and the Florida Council on Arts and Culture; Miami-Dade County Department of Cultural Affairs and the Cultural Affairs Council, the Miami-Dade County Mayor and Board of County Commissioners; and City of Miami Beach, Cultural Affairs Program, Cultural Arts Council.

Knight Foundation

MIAMIBEACH

Cover image (detail)
"Aus dem Federnwerk: Wickeln von Flugmotoren-Ventilfedern" (From the Spring Works: Winding of Aircraft Engine Valve Springs), from *50 Jahre Poldihütte* (50 Years Poldihütte), 1939.
The Wolfsonian–FIU

Inside cover (detail)
"Record of Eleven Hundred Fires," from *Modern Opera Houses and Theatres*, vol. 3, 1898.
Edwin O. Sachs (1870–1919), author.
The Wolfsonian–FIU

Published by The Wolfsonian–Florida International University, Miami Beach, on the occasion of the exhibition *Margin of Error*, November 13, 2015–May 8, 2016.

Printed and bound in the United States of America by Shapco, Minneapolis
First edition

ISBN: 978-0-9677359-9-3
ISSN: 2330-8915

Library of Congress Cataloging-in-Publication Data

Names: Abess, Matthew, author. | Wolfsonian-Florida International University.
Title: Margin of error / by Matthew Abess.
Description: First edition. | Miami Beach, FL : The Wolfsonian--Florida International University, [2015] | "No. 3 in a series of publications focused on core themes in the Wolfsonian's collection." | "ISSN: 2330-8915." | Catalog of an exhibition held November 13, 2015-May 08, 2016. | Includes bibliographical references.
Identifiers: LCCN 2015037514 | ISBN 9780967735993 (alk. paper)
Subjects: LCSH: System failures--History--Exhibitions. | Technology and civilization--Exhibitions.
Classification: LCC TA169.5 .A25 2015 | DDC 303.48/3--dc23
LC record available at http://lccn.loc.gov/2015037514

CONTENTS

Foreword

Throughout the latter half of the nineteenth century and the first half of the twentieth, the world became enthralled by new technologies that were appreciated not only for what they could accomplish, but also for what they seemed to prove—that tools separated humans from animals and firmly placed people on an imagined road to progress. Triumphal language infiltrated the rhetoric of this period and became a leitmotif that informed the modern experience. But each new invention was met with the reality of the related disasters, accidents, and unwanted complications such innovations introduced. Electricity that illuminated could also electrocute; automobiles that carried people to new places sometimes crashed; engines that harnessed power had a tendency to explode.

How did society both affirm the narrative of progress via technology and simultaneously warn people of its potentially destructive power? This question is at the core of the exhibition *Margin of Error*, organized by Wolfsonian–Florida International University curator Matthew Abess. This companion book, the third installment in a series of publications focused on The Wolfsonian's collection, features Abess's provocative essay "Margin of Error: Fields of Vision in the Industrial Age," in which he interrogates unintentional slippages found in the narrative of advancement. "What man makes, makes man," an oft-cited statement issued by the museum's founder, Mitchell Wolfson, Jr., could be reconstructed in light of this text as "what man makes, destroys man."

Those steeped in the study of modernism and its rhetoric of progress will find this exhibition and book a welcome corrective. The shadow world of hazards that was present alongside the brilliant light of progress has been minimized in the endless retelling of the dynamics of modernism. But underlying the story of the conquest of nature by machines, there coexisted warning signals. Perhaps now, in the twenty-first century, an age transfixed by airplane disasters, train derailments, and chemical spills, we can better see both sides of the coin.

The Wolfsonian is a unique museum that challenges the way we view both the past and the present. Our permanent collection of over 150,000 objects created between 1851 and 1945 allows us to reexamine the modern period and resurrect forgotten perspectives, retell in original ways stories often told, and challenge contemporary habits of mind with wisdom from the past.

For this exhibition and publication we wish to thank Micky Wolfson for his remarkable foresight and collecting genius. He has devoted his life to gathering, preserving, and sharing the objects that remain from the modern world so that we can better appreciate the power of art and design to seduce, challenge, motivate, enlarge, and delimit. Exhibitions and books such as this would not be possible without his generosity and vision.

Projects of this scope require the hard work of a multitude of skilled and dedicated individuals. The Wolfsonian is lucky to have many such people on its staff, more than I can mention by name in this space. They all have my genuine appreciation for their unique contributions to *Margin of Error*.

Florida International University has been a significant partner in all endeavors undertaken at the museum. FIU's educational mission and international vision dovetail seamlessly with Micky's original desire to create an institution guided by the highest principles of scholarship, research, and education.

Finally, I wish to express my gratitude to the individuals who have generously given their financial support to this project: Sari and Arthur Agatston, Mr. and Mrs. Henry P. Johnson, Petra and Stephen Levin, and those who wish to remain anonymous.

Timothy Rodgers
Director, The Wolfsonian–Florida International University

Margin of Error: Fields of Vision in the Industrial Age

Matthew Abess

The narratives of industrial modernity that emerged in the nineteenth century rely on a language of visual understanding. The figurative dimension of sight as a metaphor for human knowledge informed the vocabulary of mechanical improvement, identifying the eighteenth-century pursuit of philosophical clarity with the products of applied engineering. This essay considers a series of instances in which visionary promise is unsettled by practical limits and unforeseen outcomes. Episodic in its progression, the text does not chart a linear path, though each of its four sections revolves around the common theme of optical disruption as a chief characteristic of industrial times.

Beginning with the Great Exhibition of 1851 as a monument to the civilizing force of industry, the first section addresses foresight and its failures through the image of fire—not a distinctly modern phenomenon, though nevertheless an inveterate tormentor of the modern built environment, from fairgrounds to glass houses. The focus then turns to invisible currents of electricity and some of the varied means of visualizing their unseen presence, with an eye toward wonder as well as menace. This is followed by an examination of visual interference on the modern motorway and strategies for mitigating its fatal consequences. The essay concludes with a survey of the iconography of injury, centering on four case studies of state-sponsored health and safety campaigns that deployed graphic techniques to make hidden threats visible.

I. Foresight and Fire

Civilizations are built by those endowed with the abilities to anticipate and engineer, conceive and construct—such was the thesis of the Great Exhibition, London, 1851. There the nations of the world assembled to display their industries and put forward a picture of civil society in ascent. Its aspirational spirit materialized most visibly in the Crystal Palace, a colossal structure of iron and glass built to house the entire contents of the exhibition. Designed by Joseph Paxton—whose gardening work for the Duke of Devonshire included innovations in greenhouse architecture—the building consisted of modular, prefabricated components that were assembled on site in under ten months. The success of this inventiveness and economy, coupled with the building's scale, seemed to confirm that mastery of the industrial arts might bend the limits of the given (whether by God, or Nature, or Sir Isaac Newton) world.

The ordering principles of the fair reflected its organizing committee's belief in free market competition and international exchange as bulwarks of lasting peace, and in production as a measure of human achievement. Displays were divided into four classifications—Raw Materials, Machinery, Manufactures, and Fine Arts—wherein the transition from the first to the last conveyed visitors along a pageant of progress. The crude products of the earth appeared at the periphery of each national display, followed by mechanical agents of transformation, then culminating with the refined articles of manufacture and art that nature could never hand over on its own—the highest expression of human ingenuity having claimed its place in creation.[pl. 1] As the *Athenaeum* pronounced, that "distance between the raw material and the perfected work is the measure of the conquest of man over the external world—the record of that victory."[1] Emphasizing power and production, this tale of triumph made little mention of such scientific fields as electricity and magnetism, which the fair's official catalogue classified as branches of "experimental philosophy," and the tools thereof as "philosophical instruments."[2] Given that these emerging disciplines often materialized as mere parlor tricks—flashes of light in a globe of rarefied air; the charlatanism of Franz Mesmer—setting them at the margins of industrial life fit with the prioritization of proven rather than anticipated value.

In one curious appraisal of the Great Exhibition—written before the fair's official inauguration and inflected by the author's resentment over his exclusion from display—inventor and impresario Charles Babbage elaborated on the root of this mid-Victorian impulse to apprehend all worldly phenomena and mobilize them toward productive ends.[3] He argued that the "construction and the contrivance of tools, must . . . ever stand at the head of the industrial arts." Part of a wider line of reasoning, this statement exceeded pragmatic interests, as it rested on a suggestion concerning the fundamental disposition of the human race: "It is not a bad definition of *man* to describe him as a *tool-making animal*."[4] With these words the author invoked by chance a salient and ancient origin myth that also sets technics at the heart of the human character. This is the story of Prometheus, in which every step forward brings the tool-bearers nearer to their fall.

Prometheus (from the ancient Greek *promētheia*, meaning anticipatory knowledge or foresight) granted to the human race fire stolen from the hearth of Hephaestus and technical competence taken from Athena. This act was meant to be a remedy for the error committed by his brother Epimetheus (from *ēpimētheia*: reflective knowledge or hindsight), who, when distributing essences at the dawn of mortal life, neglected to grant humans any defining quality for ensuring their survival in the world to come.[5] Fire was never meant for human hands, though it was stolen on their behalf—and it is not just Prometheus who suffers for this theft. A requisite to the success of the civilizing impulse, heavenly flames elude human control. *Promētheia*, awareness of what lies ahead: despite the promise handed down by that name, experience often finds the Titan's heirs reckoning with what happened in retrospect.

Given fire's propensity for destruction, it is either prophetic or bitterly ironic that designer Henri Bellery-Desfontaines, in his poster for the 1910 Brussels Exposition Universelle et Internationale (Universal and International Exposition), expressed the fair's constructive attitude through a Hephaestus-like figure, clothed in a blacksmith's apron, bearing the tools of that trade, and proudly admiring his creation.[pl. 2] An ethereal woman floats above him, her right hand touching his shoulder while the left reaches toward the fairgrounds holding an olive branch, a symbol of peace as well as an attribute of Athena. Here then are the two victims of Prometheus's transgression—the blacksmith of Olympus and the goddess of technical wisdom—watching over a microcosm of civilization that five months later would find itself in flames. On the evening and into the morning of August 14–15, the British and Belgian sections at the center of

fig. 1 Postcard, *Panorama de Bruxelles-Kermesse après les ravages du feu* (Panorama of Brussels Fair after the Ravages of Fire), 1910

fig. 2 Postcard, *Bruxelles-Kermesse. Vers la rue de l'Escalier* (Brussels Fair: Toward the Rue de l'Escalier), 1910

the fairgrounds were completely destroyed when an ill-equipped and disorganized fire brigade failed to contain a blaze.[figs. 1–2]

Poets and prophets have long regarded fire as an agent of revelation and renewal, and it is this tradition that informs decorative artist Gordon Mitchell Forsyth's hopeful response to the destruction of August 1910 as articulated in a sizable commemorative vase.[pl. 3] Forsyth joined Pilkington's Tile and Pottery Company in 1906, where he excelled in the decorative application of rich luster glazes for the pottery's line of Royal Lancastrian wares; here this element lends a jeweled appearance to the vessel's vivid red and orange flames. Warped and tangled metal follows the contours of the fire to complete the ruinous mise-en-scène. The commemoration takes a redemptive turn, however, in the meeting of two female characters within the field of destruction: on the left, Britannia, wearing the crest of a Roman legionary though lacking her traditional spear and shield; on the right, a muse of the arts with downcast eyes, her shoulder receiving the conciliatory touch of her counterpart.[pl. 4] The two hold hands in mutual resilience, their postures poised to rise above the inferno. Art and civilization—or art *as* civilization—emerge renewed from the ashes.

Following London's example of 1851, Munich inaugurated its own Glaspalast (Glass Palace) exhibition hall on the site of its Old Botanical Garden with the 1854 Allgemeine Deutsche Industrieausstellung (General German Industrial Exhibition).[6] While this and other displays of the mechanical arts had the greatest impact on Bavaria's economic development, the annual fine art exhibitions mounted in the Glaspalast ranked highest in point of pride.[7] Among these was the 1931 Münchener Kunstausstellung (Munich Art Exhibition), which opened to the public on June 1 and ended five days later when the building was destroyed by fire.[8] Nearly three thousand works of art were lost in the conflagration, including examples by the most prominent contemporary artists of the region. Yet it was the destruction of 110 works by such masters of German romantic painting as Karl Blechen, Caspar David Friedrich, Carl Rottmann, and Philipp Otto Runge that elicited the most audible angst.[9] A statement issued by the Glaspalast-Künstlerhilfe (Glass Palace Artist Assistance) warned of the threat to the "spiritual well-spring" and "innermost life of the nation [*Volkes*]" brought on by the loss of its most distinguished arts.[10] This rhetoric finds its parallel in graphic artist Ludwig Hohlwein's poster, reproduced on stationary and stickers, for the organization's fundraising campaign to benefit those affected by the catastrophe.[11] [pl. 5] A sculptural portrait occupies the center of the image, its impassive countenance and flowing locks recalling classical depictions of Apollo, the Greco-Roman god of poetry and prophecy. The plumed crest of a Roman legionary, as with Forsyth's depiction of Britannia, renders him an emblem of endurance and rebirth.[12]

Among the many unlamented casualties of the 1931 fire was Edgar Ende's *Die Verschollenen* (The Missing), which the artist recreated in a slightly modified form in 1934.[13] [pl. 6] Static and planar, anticipating nothing, the scene sits well outside the Promethean cycle of regeneration and ascent. Ten figures (one deceased), a makeshift flag or sail, and a classical jug occupy a broken bit of land floating in ceaseless space. A grid that describes a clear line of perspective, its vanishing point recessed over the horizon, delineates the pictorial field. The rational value of such geometric construction—a nod to the orderly disposition of Italian Renaissance painting—is undercut by the dreamlike order and spatial manipulations that owe much to the *Pittura metafisica* (Metaphysical art) of Giorgio de Chirico and Mario Sironi. The scenography is also a clear reference to *Le Radeau de la Méduse* (The Raft of the Medusa), French artist Théodore Géricault's nineteenth-century masterpiece of romantic painting that depicts survivors of an 1816 shipwreck on the verge of rescue. Ende reenvisioned this historical event as a thoroughly existential catastrophe that admits neither reconciliation nor repair. There is no ship mast mounting the horizon, no definite past nor any hope of a future, only an interminable present in which lost souls drift on misshapen ground.[14]

fig. 3 View of the Crystal Palace at Sydenham, from *The Crystal Palace to Be Sold by Auction . . .*, 1911

Five years after the destruction of the Glaspalast, the Crystal Palace met a similar fate. As its residence in Hyde Park had been limited to the duration of the Great Exhibition, plans were made to relocate the building on conclusion of the fair. Reconstruction began on a new site in Sydenham in 1852 and concluded two years later with a round of ceremonies presided over by Queen Victoria. The Palace continued to host exhibitions, festivals, performances, and other assorted theme attractions, all the while weathering harsh natural elements, a stampeding elephant, and the fiery demise of its north transept in 1866. Financial woes brought bankruptcy, and the structure was put up for auction in 1911.[fig. 3] It eventually entered a public trust under whose patrimony it remained until the evening of November 30, 1936, when a small fire in the women's cloakroom grew to enormous proportions—the airiness of the structure, once a thing of wonder, now helped quicken its end.[15]

II. Invisible Currents

To press a button—*presser sur un bouton*—is "the magic gesture of modern and future stories."[16] It was with such a gesture that President Grover Cleveland set in motion the immense dynamos[fig. 4] and innumerable gears that powered the 1893 Chicago World's Columbian Exposition. In this particular story, documented with much enthusiasm by the American press, the commander-in-chief reached his hand toward the "magic key"—in fact a rather routine Victor telegraph key, though with a button of ivory substituted for rubber and a casement of gold instead of steel—and with a single touch "cause[d] an inert world to spring into animation."[17] The absence of any visible link between this modest act and its profound effects made the spectacle seem nothing short of miraculous. As reported in the *Salt Lake Herald*, it was "known in a vague way that the president was to press a golden key . . . but no one realized how intricate was this machinery, how infinite the ramifications of that electric spark." Invisible currents set miles of shafting into revolution as flags unfurled, fountains sprang to life, and over one hundred thousand incandescent lights announced the arrival of the electric age.[18]

Wonders overflowed from the exposition's Palace of Mechanical Arts—motors, printing presses, dental devices, an all-electric kitchen range—where an annex of Westinghouse generators fed with oil from underground pipes produced enough electricity to service the entire fair.[19] Among the most dazzling of these marvels were the colossal searchlights that shined over the Court of Honor each night, frequently cited as metaphors for clarity and truth. Yet if the stereographic camera-eye even somewhat approximates lived experience, then by all appearances light roundly overwhelmed the to-be-enlightened visual field.[fig. 5] Their blinding brilliance recalls the painful, if eventually redemptive illumination that assaults the prisoner in Plato's allegory of the cave; on the "upward journey of the soul to the intelligible realm,"[20] he emerges from a darkened world, and there must avert his eyes from the glaring light to take refuge among reflections and shadows. Suffuse and ineffable, seen only in its effects—and sometimes not even then—electricity poses a representational dilemma.

"To press a button"—*presser sur un bouton*—"will be looked upon as a very archaic and very strange method of sorcery."[21] So suggests *Électricité* (Electricity), a 1931 portfolio of ten photographic images

fig. 4 Stereograph, *The Immense Dynamos, Machinery Hall, Columbian Exposition*, 1893. B. W. (Benjamin West) Kilburn (1827–1909), photographer

fig. 5 Stereograph, *Illumination and Great Search Light, Columbian Exposition*, 1894. B. W. (Benjamin West) Kilburn (1827–1909), photographer

created by Man Ray on commission from a private utilities company in Paris as part of its campaign to encourage domestic use of electricity.[22] Constructed almost exclusively without the use of a camera—most significantly, through the direct exposure of objects to light-sensitive paper—the sequence puts forward one means of casting electrical apparitions into the visible realm: abandon the button altogether. It is to this relinquishment that author Pierre Bost alludes in his introductory text when he describes the "magic gesture" of turning a switch and the "archaic" air it has assumed owing to Man Ray's largely cameraless (hence, buttonless) treatment of the subject. The technique of creating images without, or with minimal intervention of, a lens was especially suited to the task of giving visible form to electrical phenomena; the darkroom's electric light traces shadows of objects onto paper, thereby inscribing its source of power into the optical field. While not a direct exhibition of invisible force—that would be impossible—the largely unmediated nature of the exposures foregrounds a forceful presence.[23]

The images that Man Ray produced document the numerous and astonishing functions performed by this mysterious medium. *Le monde* (The World)[pl. 7] sets an electrical cord—complete with a button—beneath an enormous moon in a starless sky, such that illumination of the celestial spheres becomes the province of a modest switch. The plate titled *Électricité* [pl. 8] likewise invokes astral bodies in the form of a lightbulb keeping tender watch over the stars; following Bost's oblique narrative—"We want to see tomorrow in the glittering array of stars, as yesterday (the day before yesterday) we saw the spilled milk of Hera"[24]—the lightbulb is akin to the ancient Greek queen of the gods, whose spilled breast milk was said to be the origin of the Milky Way.

Electricity animates the Parisian landscape in *La ville* (The City),[pl. 9] where a vertiginous display of fragmented neon texts encircles and arouses the Eiffel Tower. *Salle à manger* (Dining Room)[fig. 6] is a pure photogram (or "rayograph," as Man Ray preferred), produced using only an electric toaster placed on photographic paper and exposed to electric light.[25] The appliance hovers in pictorial space, a phantasmagoria of comfort and convenience in the home, revealing the commercial alliance of electric utility companies and appliance manufacturers as together they marketed to middle-class housewives the virtues of a labor-saving, all-electric home. The cameraless exposure coupled with the openwork form of the toaster—out of fashion for some years, and notably less safe than its enclosed counterpart—exposes the appliance's inner coils to construct an anatomical portrait. Notions of the body appear as well in *Lingerie,*[fig. 7] an image of an electric iron together with some object that resembles distended digits; the fetishistic title and downward thrust of the appliance at the hand of a disembodied agent bring to the fore forces of attraction, whether erotic, violent, or both. A second image assigned the title *Électricité*[pl. 10] unites femininity (the nude torso of Lee Miller posed after the *Venus de Milo*) and the excitation of electrical currents (wavy lines that cut diagonally across the frame) to conjure a "dangerous goddess" harboring "dangerous secrets"[26]—silent, seductive, capricious, and unsparing.

Associations with femininity appear throughout the history of electrical science, in which the wondrous force has been variously characterized as female fire and electric fairy. In Adolfo Hohenstein's poster for the 1899 Esposizione Internazionale di Elettricità e Esposizione

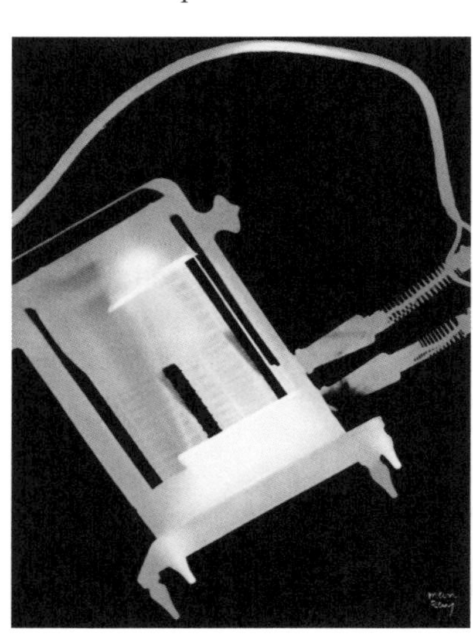

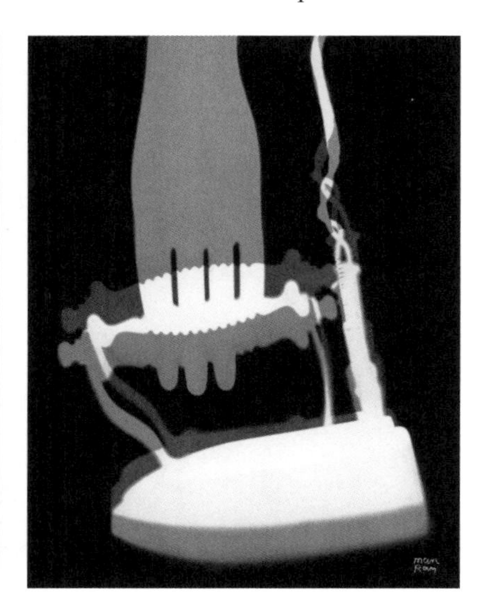

fig. 7 *Lingerie*, 1931. Man Ray (1890–1976), photographer

Nazionale Serica (International Exposition of Electricity and National Exposition of the Silk Industry) in Como, Italy,[pl. 11] electrical posts are drawn with equal line weight as the two female figures in the foreground of the image, whose arms and hands are posed with the same postural grace as the power lines behind them. Typical of the Liberty style, of which Hohenstein was an exemplar, the entire scene is framed within a sensuous floral border. One of the women holds a shuttlecock and wears flowers in her hair; the other rests one hand on what is either a weaving reed or a voltaic pile, while with the other hand she places a wreath around a cameo of Alessandro Volta, namesake of the unit of electromotive force. The city of Como was eager to promote its excellence in the silk industry; that Volta hailed from there, and that the centennial of his discovery of the battery coincided with coal-poor Italy gaining pace in the utilization of waterpower, provided sufficient pretext for mounting an exhibition in his honor. Reports from the time tend to concur that the electricity section of the exhibition was less compelling than expected; the power looms and products of the silk section seem to have been well received. The joining of the two in such near proximity turned out to be disastrous, as an errant spark from an electrical machine came into contact with one of the silk displays, razing the exhibition just six weeks after its opening. Here again, the dangerous goddess, overshadowing Hohenstein's pictorial conflation of feminine idyll with electrical life.

fig. 8 Book cover, *Die Berliner Elektrizitätswerke bis Ende 1896* (The Berlin Electric Power Stations

The cover of a rather corporate volume extoling Berlin's electric power stations in the year 1896[fig. 8] bears a decidedly more peculiar stamp of gendered forces in the modern world. The image, by designer Ludwig Sütterlin, sets a nude man over billowing flames as the personification of a steam boiler. With his right hand he grips the wrist of a woman at whom he forcefully exhales a cloud of steam, which strikes her as if she were the turbine of an electrical generator. Jagged lines dart out in all directions from her fingertips, though by her laconic gaze, she appears to take little notice. The association of masculinity with muscular force and of femininity with passive, atmospheric flows is depicted here in no uncertain terms.

The alliance of divinity and electricity, meanwhile, predominates in images where electrical force is denoted by incandescent light. Daan Hoeksema's winning poster design for the 1907 Tentoonstelling van Electriciteit in Huis en Ambacht (Exhibition of Electricity in the Home and Handwork) in Arnhem, the Netherlands [pl. 12]—a show of electrical applications and appliances for industrial and domestic use, mounted to coincide with the unveiling of the city's municipal power station—substitutes an incandescent lamp for a church bell, ringing in the modern moment for the silhouetted old town in the background.[27] The call for design submissions mandated a three-color scheme that invoked at least one of the following motifs: light and power, strength, improvement, amps, unity, and silent force.[28] Of the thirty-nine proposals taken under consideration, Hoeksema's was selected for its artful interpretation of the theme, wedding sober deference to tradition with the promise of bright days ahead.

An advertising image for the Osram incandescent lightbulb [fig. 9] is even more audacious in its conflation of heavenly power and artificial illumination. Not quite at the center of the solar system, an Osram lamp box intervenes between the sun and humankind, saturating the Earth with its vivifying rays. Photographer John Havinden—founder and principle of Gretton Studios, one of the most prominent commercial photography firms in Great Britain—created the image around the time of the 1936 International Surrealist Exhibition in London. The formal influence of that movement shows clearly in the composition. Yet where Surrealism aimed to reconfigure consciousness—and with that the prevailing sociopsychological order—Havinden's photograph gives image to the popular commercial refrain that the twenty-four-hour radiance of incandescent light signified the triumph of technics over the diurnal limits of the sun.

The ancient gods likewise make frequent appearances in the representational canon of electrical force. The winged helmet of Mercury speeds currents across a seven-mile span between the eastern border of the Netherlands and the German region of North Rhine-Westphalia; [pl. 13] a formidable hero wrestles an unruly bolt into useful submission in an image for the 1922 Electrical and Industrial Exhibition in Manhattan, [pl. 14] where the one jumbo

fig. 9 Photograph, *Osram*, c. 1935, John Havinden (1908–1987)

dynamo not destroyed in the Pearl Street Station fire of 1890 was put proudly on display in commemoration of forty years of Edison service in New York;[29] and a Promethean character hovers above a generator and catches lightning from the clouds for the benefit of humankind as the representation of the 1928 Electrical Industrial Exhibition,[pl. 15] with nary an indication that the great number of wares on view were in the field of electrotherapeutic devices—UV lamps, vibrators, massage chairs, magnetic rejuvenators, and mechanical camels "guaranteed to 'duplicate the motion of the ship of the desert' and said to be especially good for the liver."[30] That such imagery seems more elevated than the events it describes indicates the degree to which even a detoxifying quadruped was seen as an expression of the magnitude of human victory over darkness and disarray.

Mythic scale is also central to the narrative of electrification as told by Westinghouse Electric and Manufacturing Company at the 1933 Chicago A Century of Progress International Exposition,[fig. 10] where six panels illustrated the company's significant contributions to advancing modern industry and American eminence: alternating current transmission and its varied applications, such as servicing the 1893 exposition; harnessing hydroelectric power from Niagara Falls; innovations in steam turbines and turbo-generators; commercial radio broadcasting; and high-voltage railway electrification. The last of these "Westinghouse Pioneering" panels[pl. 16] centers on a colossal personification of high voltage carrying electrical charge from a generative rotor (denoted by a circle at his feet) to the high-tension towers that rise to meet his hands. Trains dart out from either side, their force of motion magnified by the sweep of semicircular forms and the staccato rhythm of the mountains pressing outward from the center. The frieze at the top, with an agricultural landscape on its left and industry on its right, announces 1905 as the year when electricity united the divided expanse of the nation. It is an ennobling scene that abstracts the unseen drama of electricity into visually arresting, familiar forms.[31] It is also a thoroughly discomfiting composition, conjuring the hazards of electrified bodies and the danger of the third rail.

To press a button—*presser sur un bouton*—"[its] wonderful force would certainly appall any but a god."[32] When capital punishment entered the purview of the electrical field, it was boasted that, in the near future, one might simply press a button from the safety of a legislative office and thereby boil a body many miles away. Through the efforts of Thomas Edison, the spectacle of remedial electrocution in fact played a significant role in fostering public awareness of electricity's many perils.[33] Still, there were only so many occasions for such displays, while the hazards of this hidden force grew ever more pervasive. Pressure to safeguard the populace against the menace of errant currents gave rise to a new repertoire of visual motifs. The electric fairy of yore might take the form of a phantasmal creature lurking on exposed wires,[pl. 17] or a wide-eyed serpent running from knob and tube to frayed nerve.[pl. 18]

A poster issued by the Austrian Central Office of Accident Prevention provides an evocative and informational account of the dangers that might assault the human body as it endeavors to change a

fig. 10 View of the Westinghouse Electric and Manufacturing Company display at the 1933 Chicago A Century of Progress Exposition

lightbulb.[pl. 19] The psychological impact of Joseph Binder's design pivots on its image of man as a conductive organism, disfigured in light of a careless touch;[34] the accompanying text offers detailed instructions on the dos and don'ts of the precariously unfamiliar act. The logo of the larger work safety initiative to which the poster belongs (imprinted on the lower right) is a vigilant eye accompanied by the admonishment, "Gib acht!" (Be Careful!).[35] The same motif of fatal bolt and jolted body appears in a poster from the Finnish Federation of Accident Insurance Companies, most likely published in conjunction with a traveling national exhibition on occupational health and safety.[pl. 20] A misplaced thumb precipitates thunderous contact between a charged bulb and the cage of a lamp guard, letting lightning loose through the formerly productive right hand as its spectral copy reaches into the frame and clasps a skeletal counterpart. The text beneath this scene transmits a most memorable equivalence: "A damaged hand lamp is a deadly handshake." Bridled and practical in certain hands, of fatal consequence in others, the beneficent fairy and dangerous goddess of electricity are never without the other.

III. Automotive Safety

The story of terrestrial transportation as it evolved in the early twentieth century has tended in its telling to follow an equine progression: bipedal motion gave way to the saddled horse, which was put to work drawing chariots and cabs until superseded over the run of centuries by the locomotive (as "iron horse") and the automobile (or "horseless carriage"). According to the General Motors Department of Public Relations, this incremental disappearance of the horse corresponded precisely with one of the chief quandaries of the modern motorist—obstructed vision:

> When our father climbed from a horse and buggy into a horseless carriage, our problems in vision increased. A very valuable pair of eyes had been sacrificed. No longer could we rely on old Dobbin to help guide the shay—we perform the task ourselves, and because of the inherent nature of our vehicle we see a good deal less of the road than did our faithful horse.[36]

New spheres of motion hatched unforeseen obstacles—mechanical, cognitive, and physiological—that the General Motors Corporation set out to disperse. Invoking Paul the Apostle's first letter to the Thessalonians—"Prove all things; hold fast to that which is good"—it was to this end that in 1924 the company established its 1,268-acre General Motors Proving Ground in Milford, Michigan, "a small, isolated world where the Division engineers, unhindered by countless traffic regulations and without danger to the general public, can perform any tests they desire under all driving conditions."[37] Areas of inquiry included the maximum allowable stress in coiled springs; noise abatement; brake endurance; steering tension; and the ideal center of gravity, which was calculated with reference to Galileo when one Proving Ground scientist recalled the law of the pendulum, thereby "adapting a three hundred year old discovery" to answer "an important question of today."[38] As for sight, scientists labored to optimize the motorist's field of vision using an illuminated automobile interior that cast shadows on a checkered wall;[fig. 11] each silhouette corresponded with a source of visual interference that it was the technician's charge to eliminate. This was an especially crucial element of comfort and safety in those instances when vision was deformed by the vagaries of night—a condition further exacerbated by the introduction of high-luminosity incandescent headlights.[pl. 21–22]

Seeing was of little use, however, absent standardization of roadway behavior. Regulatory authorities presided over the systematic installation of directional signage across the expanding automotive environment, with omnipresent roadside markers guiding both drivers and pedestrians through the blurry domain of safe conduct. The prevalence of youth casualties was an especially vexing concern. Walt Disney's exasperation at the failure of his Autopia attraction, conceived as a microcosm of civil order, to inspire spontaneous adherence to the social contract—the subjects instead took great delight in smashing their bumpered cars into those of their peers—reflects the very real problem presented by enthusiastic youth untutored in the dangers of vehicular life. Schools in the Netherlands addressed this issue in part by incorporating grade-level railway and four-way intersection markers into their *Teekenen en kleuren* (Drawing and Colors) lesson plan for primary school children grades one through four.[pl. 23–24] The easily read reductive geometry and bold, minimal coloring of the images would have served as a useful primer in the harmless cohabitation of vulnerable bodies and fast-moving machines. Such clarity was of paramount concern in the management of the chaotic modern roadway, where interpretation and swift response were easily submerged in a barrage of visual instruction.[pl. 25] Civic agencies worked closely with urban planners and designers to devise efficient schemes for protecting their citizens from the hazards of the road. In the Dutch township of Schoonhoven, city architect Died Visser turned to the expressive principles of the De Stijl movement—precise geometric construction and a preference for primary colors—in his proposals for traffic signals,[pl. 26–27] carrying the utopic ideals of progressive artists into the practical field of municipal design.

VISIBILITY FROM
DRIVERS POSITION

The adoption of uniform governance rarely kept pace with the growing complexity of transportation networks; particularly where old and new means of conveyance shared infrastructure not yet adjusted for coexistence, misfortune predictably followed. This was the case for nearly a century on Manhattan's congested West Side, a harrowing corridor of injuries and fatalities where an especially hazardous Eleventh Avenue was widely known as "Death Avenue." The thoroughfare comprised an intricate web of intersecting traffic: goods traveling by carriage or car to and from the busy Hudson River piers; children coming and going from schools and recreation centers; everyday travelers on horse or on foot; and, most dangerously, the long trains of the New York Central freight line that carried tons of food into the city along grade-level tracks. The situation was worsened by population density, as waves of immigrants disembarked at the city's ports and settled into tenements centered around the tracks. The only protective measure in place was an 1850s ordinance mandating that all trains be preceded by a man on horseback carrying a red flag and riding at a pace not to exceed six miles per hour. These mounted signalmen (or "West Side cowboys") were not entirely effective. Over its decades of use, the street-level line of the New York Central was responsible for hundreds of deaths, whether by maiming, flattening, or decapitation, as well as thousands of injuries both minor and severe. Most of the victims were children, with the fatality numbers rising in the dark and slippery winter months. It was the killing of a seven-year-old boy in 1908, whom the *New York Times* described as having been "ground to death,"[39] that incited citizens to agitate in earnest for removal of the tracks. A League to End Death Avenue was formed, with ministers of West Side churches lobbying from the pulpit during Sunday Mass for the rail line to be submerged.

The West Side Improvement project was finally initiated in 1929 as a two-pronged solution to the blight of Death Avenue: removing trains from street level to an elevated rail that provided direct warehouse access for off-loading meat and produce (now the location of the much-celebrated High Line park) and the construction of an elevated motorway along the Hudson River for relieving automobile traffic congestion. A photograph by Gordon Coster, *Silhouette – N.Y.C. Elevated Motor West Side under Construction near 14th St.*, shows the progress of the motorway project as of February 1930.[fig 12] Three men stand on the nascent infrastructure, and what appears to be smoke from a locomotive can be seen rising in the background. Coster captured his industrial subject with an attention to formal properties, with the scaffolding and cranes serving to frame the hazy sky. Poor engineering and delayed maintenance plagued the once-innovative

fig. 11 "Eyes that Cast Shadows," from *Putting Progress through Its Paces: The Story of the General Motors Proving Grounds*, 1937. Otto Linstead (1887–1970), photographer

fig. 12 Photograph, *Silhouette – N.Y.C. Elevated Motor West Side under Construction near 14th St.*, 1930. Gordon Coster (1906–1988)

structure from the start; it was finally closed in the early 1970s after a tractor-trailer plunged through its disintegrating surface.

Should the instructional efforts of municipal offices or the rehabilitation of public works fail to come through on some darkened road, there remains as recourse only the highly developed instinct for survival. The efficacy of this last resort is conveyed in a dramatic diagram by Herbert Bayer [pl. 28] for the July 24, 1939, issue of *Life* magazine, where it accompanied an article covering neuroscientist James W. Papez's research on brain anatomy and its relationship to mechanisms of thought and emotion. In consultation with Papez, Bayer mapped the cerebral circuits that lead from consciousness ("the brain's waking mechanism") and sensation ("a conscious experience, aroused by impulses from the sense organs") to perception ("the vital act of recognition") and voluntary action ("carried out through the downward connection, to the spinal cord, motor nerves and, finally, the muscles").[40] This complex process of comprehension and response is illustrated with the example of oncoming headlights: the man senses them with his eyes, registers them as a danger, and swiftly removes himself from the path of collision—all within a fraction of a second. The novelty of the sensation does not prevent it from penetrating the premodern regions of the mind, where reflexive excitation of neural pathways may yet give hope to accident statisticians dispirited by their mounting data.

IV. Industrial Hazards

As the primary means by which humans take hold of and act on the world, hands have long been raised as an emblem of creative force.[fig. 13] Their frequent appearance in the iconography of injury is therefore doubly devastating, as the hand that makes also becomes the agent of its own undoing. The spectacle of wounded limbs increased with the advance of modern industry, as bodies mobilized into the service of mass-production were exposed to new and indefinite dangers. Occupational safety initiatives developed in kind with the spread of mechanical risk, giving rise to a graphic language of peril and prevention, imperative and admonishment. The four brief case studies that follow offer an introduction to the informational strategies of accident awareness as they emerged in the decisive period between the two world wars. Produced by state agencies, each sequence of images manifests the particular traits of a national culture, political climate, and economic market. Common throughout is the belief that human error is among the most regular causes of injury and, consequently, that the burden to uphold standards of safety sits principally with workers.

Case Study 1: Rijksverzekeringsbank, The Netherlands[pl. 29–36]

Occupational health and safety legislation in the Netherlands began soon after the country's late entrance into the era of steam. The provisions of its prevention and compensation laws related to accidents in the workplace—at first quite limited, these laws were applied more broadly as the Dutch social welfare state solidified its positions in the period following the First World War—were written on the premise that private members of society should not bear the fiscal burden of damages borne of private contractual relationships. Employers therefore held responsibility for premiums while a government insurance agency assumed the risk. This agency, the Rijksverzekeringsbank (National Insurance Bank), financed and oversaw an additional initiative in the interest of promoting a culture of awareness—the Veiligheidsmuseum (Safety Museum), known before 1914 as the Museum van Voorwepen ter Voorkoming van Ongelukken en Zeiten in Fabrieken en Werkplaatsen (Museum of Objects to Help in the Avoidance of Accidents and Diseases in Factories and the Workplace).

The Safety Museum offered free admission in its effort to inform the greatest possible number of employers and employees of the various dangers that threaten life and limb in each major branch of industry, as well as the most pragmatic means of safeguarding

fig.13 "Aus dem Federnwerk: Wickeln von Flugmotoren-Ventilfedern" (From the Spring Works: Winding of Aircraft Engine Valve Springs), from *50 Jahre Poldihütte* (*50 Years Poldihütte*), 1939

against them. In 1922 it added to its highly active roster of activities the distribution of so-called safety images, published under the auspices of the poster committee of the National Insurance Bank. The eight posters reproduced here, many of which include an invitation to visit the Safety Museum, are emblematic of the bank's approach to the prevention of accidents in the workplace. Emphasis is given to unsafe handling and operation of machinery, placing the duty of compliance firmly with the worker. Positive images showing proper conduct do make appearances—the woman who remembers to protect her hair in a tie; the man who thinks of his mother—though the motivating force of mutilation is far more often invoked.

Case Study 2: Úrazová pojišťovna dělnická pro Čechy, Czechoslovakia [pl. 37–44]

Founded in 1889 as Úrazové pojišťovně pro Království České (Accident Insurance Agency of the Kingdom of Bohemia) and renamed Úrazová pojišťovna dělnická pro Čechy (Workers Accident Insurance Agency of Bohemia) following the dissolution of Austria-Hungary, this regional agency is best remembered as a former employer of Franz Kafka. Less well known is the series of striking accident prevention posters that it commissioned and distributed directly to factories beginning around 1934. An internal report published in 1935 described its efforts at ensuring best practices in the industrial workplace as focused on "advertising labor mishaps by means of alarming images"; lacerated hands and impending division from limbs are the most prominent motifs, followed by violently compressed bodies. Many of the designs apply the tenets of asymmetric typography to staged photographic reportage, setting forth imperatives that are both legible and impactful.

Case Study 3: Ente Nazionale di Propaganda per la Prevenzione degli Infortuni, Italy [pl. 45–56]

The Kingdom of Italy inaugurated a program of social welfare in the last decade of the nineteenth century, establishing a standard of state intervention with respect to capital and labor that was later adapted to the corporatist model of Fascist governance. While not dissimilar in scope of interest—maintenance of health and social relations remained of primary concern throughout—social welfare under the Fascist regime was carefully framed to generate political support and foster national belonging; ensuring the safety of the body politic and managing the spiritual health of its citizens held a primary place in the performance of Fascist justice and security. The positioning of the state as a parent to its people is a distinguishing feature of the Fascist approach to accident prevention. A series of work safety postcards issued in 1938 by the Ente Nazionale di Propaganda per la Prevenzione degli Infortuni (National Board of Propaganda for the Prevention of Accidents) combines vivid graphics with rhyming couplets and quatrains to remind laborers of the many *malanni inenarrabili* (unspeakable misfortunes) provoked by careless behavior.

Case Study 4: CSM, Belgium [pl. 57–60]

The Belgian mining industry of the 1920s invested heavily in increasing the production of its long-stagnant collieries. The rate of accidents in what remains one of the most dangerous occupations multiplied in turn—a situation further intensified by the rapid influx of foreign labor in 1923, turning the cacophony of the mine into an informational nightmare. An agency with the acronym CSM produced a series of multilingual warnings that could be understood by the diverse multinational community laboring in the shafts. Clear, unemotional, and occasionally misspelled statements of fact frame woodcut illustrations inflected by minimal appearances of red—a scarf, a brick wall, and splashes of blood. Each image presents a different moment on the road to misfortune: a fellow miner intervening before disaster, a body in the midst of experiencing disaster, a life ended by disaster, and a supervisor scolding a survivor of disaster for failure to wear his leather cap.

Meanwhile on the American scene, as the proliferation of skyscrapers and bridges in the mid-1930s announced a turning point in the Depression decade, images of construction workers riding hoists and

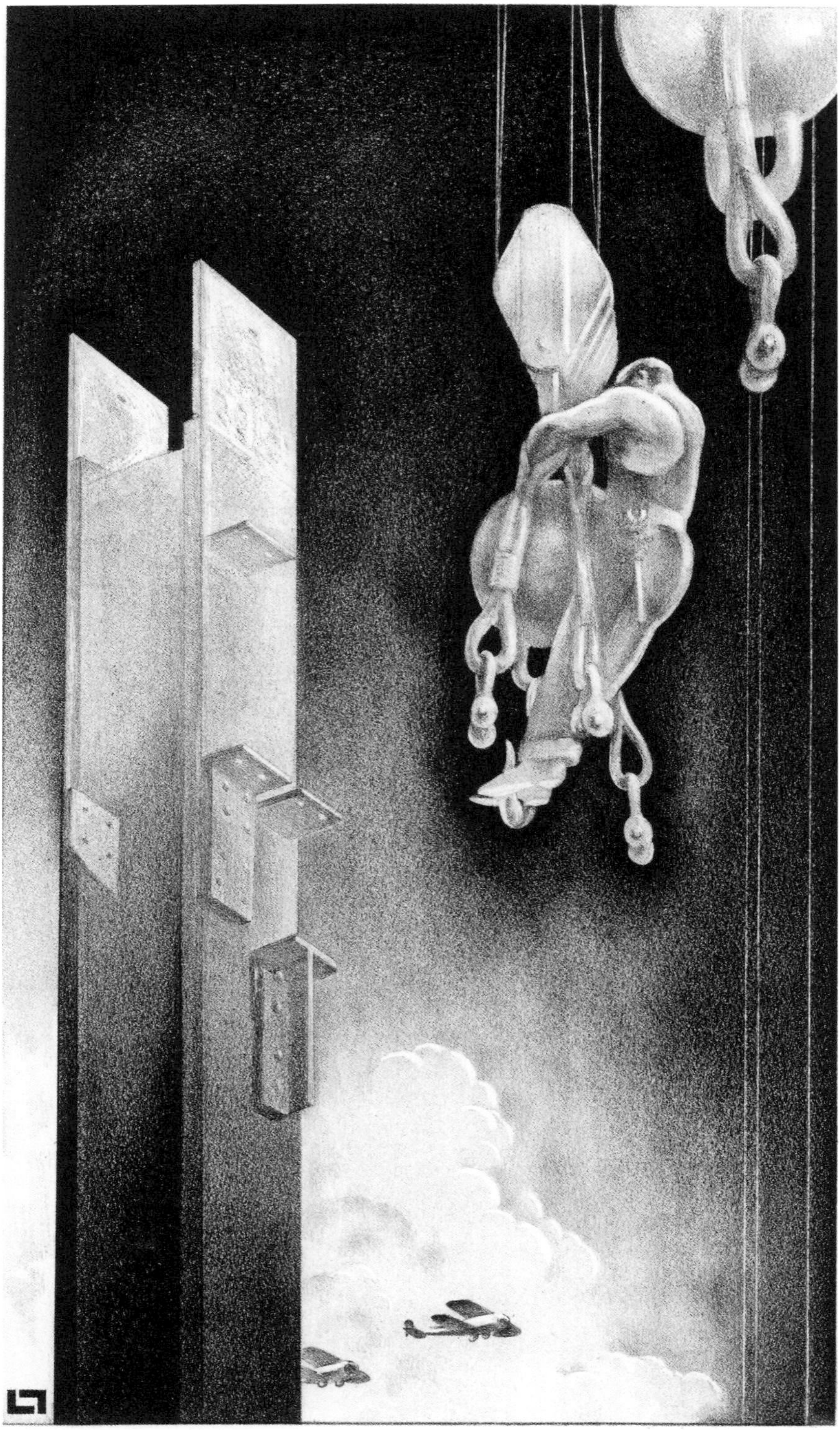

fig. 14 Print, *Mid-air*, 1931. Louis Lozowick (1892–1973)

hooks affirmed the fortitude of American will for the weary public imagination. A circa 1935 photograph by Wendell MacRae, *Nerve and Steel*, captures one of these heroic and potentially perilous moments as encountered during the construction of Rockefeller Center.[pl. 61] The worker here has none of the virile muscularity that distinguishes a large number of artistic responses to the American industrial epic; his tousled work wear and tight embrace of the ball and hook belie exhaustion and fear. Even so, he rises above the bonds of the Earth, his ascent augmented by the rhythmic play of diagonal vectors (girder, crane, and cables) that sweep him ever upward. The February 1940 issue of *Fortune* reproduced the photograph in an article titled "The Culture of Democracy," where it was captioned with the final verse of Walt Whitman's 1874 poem "Song of the Redwood-Tree":

> Fresh come, to a New World indeed, yet long prepared,
> I see the Genius of the modern, child of the real and ideal,
> Clearing the ground for broad Humanity, the true America, heir of the past so grand,
> To build a grander future.

The staunch verticality of Louis Lozowick's 1931 lithograph *Mid-air*,[fig. 14] an early example of the artist's shift from precise machine ornaments and geometric order to depictions of those who forge and construct, directly underscores the dauntless character of the native "genius" that Whitman describes.

Sociopolitical changes of the intervening years naturally subjected *Mid-air* to reinterpretation. The *Monthly Labor Review* reproduced it on the cover of its October 1983 issue; in April 1984, the following letter to the editor appeared beneath an article on productivity in meatpacking:

> [In 1931] a construction worker riding to work on the ball and hook of a crane [may have been] a symbol of the American work force—tough, fearless, and hard-working. But to the modern safety professional, this scene is another, more frightening kind of symbol. It illustrates real hazards that were common practices 50 years ago. . . . Even today, improper hoisting of personnel causes tragedies. Last spring, for example, four workers were killed while being lifted by a crane during construction of a Florida football stadium.
>
> Thorne G. Auchter
> Former Assistant Secretary of Labor for Occupational Safety and Health
> United States Department of Labor

I extend my gratitude to Jon Mogul for his comments on earlier drafts of this essay, which have strengthened it in every way; to Kara Pickman for her astute editorial feedback; to Lisa Li for her attentive proofing; and to Taylor Anne Abess, for all of this and more. Rachel Fesperman contributed invaluable research, and Peter Clericuzio helped procure some of the more elusive primary source materials. Many thanks are due to Silvia Barisione, Nicolae Harsanyi, and Eva Melnikova for their translations of foreign-language titles. Translations within the text are my own.

1 Quoted in John Tallis, *Tallis's History and Description of the Crystal Palace, and the Exhibition of the World's Industry in 1851* (London and New York: John Tallis, 1852), 1: 14.

2 See Jeffrey A. Auerbach, *The Great Exhibition of 1851: A Nation on Display* (New Haven, CT: Yale University Press, 1999), 122–27.

3 Unwavering (and by all accounts, entirely insufferable) in his pursuit of reason and utility, Babbage developed a number of mechanical aids for advancing essential industries. His "self-registering apparatus," which he claimed could record "the immediate antecedents of any catastrophe," may be considered the forebearer of the present-day black box. See Charles Babbage, *Passages from the Life of a Philosopher* (London: Longman, Green, Longman, Roberts, and Green, 1864), 329–34.

4 Charles Babbage, *The Exposition of 1851; or, Views of the Industry, the Science and the Government of England*, rev. ed. (London: John Murray, 1851), 173.

5 Plato, *Protagoras* 320d–322e, in *Plato: Complete Works*, ed. John M. Cooper (Indianapolis: Hackett, 1997), 756–58. See also Bernard Stiegler, *Technics and Time*, vol. 1, *The Fault of Epimetheus*, trans. Richard Beardsworth and George Collins (Stanford, CA: Stanford University Press, 1998), 185–203.

6 Drawing on Paxton's innovative use of prefabricated building components, architect August von Voit's plans were realized just eighty-seven days after foundation work commenced. Georg Kohlmaier and Barna von Sartory, *Houses of Glass: A Nineteenth-Century Building Type*, trans. John C. Harvey (Cambridge, MA: MIT Press, 1986), 336–43. See also Christoph Hölz, "Glaspalast," in *Zwischen Glaspalast und Maximilianeum. Architektur in Bayern zur Zeit Maximilians II, 1848–1864*, ed. Winfried Nerdinger (Eurasburg, Germany: Edition Minerva, 1997), 120–25.

7 Other prominent exhibition themes included agriculture, animal husbandry, botany, and beer.

8 The exact cause of the fire remains a matter of debate. While official investigations ruled it to be spontaneous combustion, the claim of arson remains more plausible.

9 See Georg Jacob Wolf, *Verlorene Meisterwerke deutscher Romantiker* (Munich: Bruckmann, 1931).

10 Leaflet issued by Glaspalast-Künstlerhilfe e. V. München, quoted in *Aurora: Ein romantischer Almanach* 2 (1932): 97.

11 See Kurt Pfister, "Die Glaspalast-Künstlerhilfe," *Deutsche Kunst und Dekoration* 68 (April–September 1931): 306–07.

12 In its fatalism, fetishism, and evocation of the enduring *Volk*, this purgative scene presages the National Socialist attack on experimental form as visible evidence of a nation's moral and spiritual decline. Adolf Hitler opted not to rebuild the Glaspalast, electing instead to follow a design by architect Paul Ludwig Troost for a new building at the edge of the English Garden. The plans for the so-called Neuer Glaspalast (New Glass Palace) were realized as the Haus der Deutschen Kunst (House of German Art).

13 Published references to this work are inconsistent with respect to its title. The original painting that was destroyed in the 1931 Glaspalast fire is reproduced as *Verschollen* (Missing or Forgotten) in Franz Roh, "Edgar Ende. Ein 'surrealistischer' Maler in München?," *Die Kunst* 64, no. 4 (January 1933): 122–26.

It is listed together with the 1934 version as *Auf der Scholle* (On the Soil or On the Clod of Earth) in Jörg Krichbaum, ed., *Edgar Ende: Der Maler geistiger Welten* (Stuttgart: Edition Weitbrecht, 1987). This monograph also reproduces a nearly identical image of a 1931 watercolor under the title *Die Scholle*, to which a reference also appears in *Edgar Ende. Aquarelle, Gouachen, Zeichnungen, Lithographien* (Munich: Galerie Wolfgang Ketterer, 1974). The matter is further complicated by the fact that none of these appear in the 1931 Münchener Kunstausstellung exhibition catalog, though there is broad consensus that they were on the premises of the Glaspalast at the time of the fire. The title *Die Verschollenen* (The Missing) used here in reference to the 1934 painting is taken from the 1935 exhibition catalog *Münchener Kunst: Sonderausstellungen in der Neuen Pinakothek* (Munich: Knorr & Hirth, 1935), 13, as it is the most proximate and primary source.

14 Though not a supporter of the Nazi party, Ende occupied an uneasy place in its artistic canon. His academic approach to composition and resonance with the romantic tradition made his work acceptable in theory, yet his strange and visionary scenes did not merit inclusion in the annual exhibitions of approved art at the Haus der Deutschen Kunst. See Peter Breuer, *Münchner Künstlerköpfe* (Munich: Georg D. W. Callwey, 1937), 274–76, a biographical index of Third Reich–approved artists from the Munich region.

15 There is some solace to be had in the fact that the fire preceded by some twenty-four hours the arrival of many prize cats for the city's great cat show. None therefore were harmed in the destruction of the Crystal Palace.

16 Pierre Bost, introduction to *Électricité* (Paris: Compagnie Parisienne de Distribution d'Électricité, 1931), n.p.

17 "Cleveland Presses the Golden Key," *Salt Lake Herald*, May 2, 1893.

18 The astonishment affiliated with pressing a button repeated itself at the 1900 Paris Exposition Universelle: "One gentle touch of an electrical button with the finger unbridled the magic fluid and sent it on its instantaneous career to emblazon palace, hall, tower, museum, shop and garden, and impart its energy to myriads of mechanical devices." James P. Boyd, *The Paris Exposition of 1900* (Philadelphia: P. W. Ziegler, 1900), 237.

19 Westinghouse's receipt of the contract to light the fair initiated the victory of alternating current (AC) over direct current (DC) power transmission—the crux of the so-called War of Currents that pit Thomas Edison's DC system against the AC system championed by George Westinghouse and Nikola Tesla.

20 Plato, *Republic* 517b, in *Plato: Complete Works*, 1135.

21 Bost, introduction to *Électricité*, n.p.

22 The portfolio was produced by the Compagnie Parisienne de Distribution d'Électricité (CPDE) in an edition of five hundred and distributed to executives, significant shareholders, and top clients. One can presume that its aim was to assure these parties of the company's continued commitment to expanding electrical energy consumption at a time when economic depression had reduced its industrial client base and natural gas remained both a more established and less expensive option for domestic needs such as cooking and heating. It is doubtful that the sought-after market of middle-class customers was ever much aware of its existence.

23 Compare this with what Man Ray described as the "wonderful technical photographs of the interior of powerhouses" that another photographer had previously taken for the CPDE project. Arnold Crane, Interview with Man Ray, 1968, in the Archives of American Art, Smithsonian Institution, quoted in Stefanie Spray Jandl, "Man Ray's *Électricité*," *Gastronomica* 2, no. 1 (2002): 14.

24 Bost, introduction to *Électricité*, n.p.

25 Most of the images in the series involve the use of negatives or multiple exposures in addition to cameraless elements.

26 Bost, introduction to *Électricité*, n.p.

27 Arnhem was among many Dutch cities that delayed sponsoring electrification, no doubt owing to recent investments by various municipal councils in gas plants previously under private ownership. See Pim Kooij, "'Where the Action Is.' The Introduction and Acceptance of Infrastructural Innovations in Dutch Cities, 1850–1950" (paper presented at the World Economic History Conference, Helsinki, August 25, 2006).

28 *De Opmerker* 47, no. 6 (1907): 46–47.

29 In light of the fire that brought down Edison's Pearl Street Station, it is not surprising that safety fuses for domestic use were especially well represented in the commercial displays.

30 "Electric Industry Displays its Wares," *New York Times*, October 21, 1928.

31 The sequence of panels also provided Westinghouse with an occasion to display Micarta, a new laminate the company hoped to promote for such uses as interior decoration and mass merchandising.

32 "Grover to Press the Button," *St. Paul Daily Globe*, May 1, 1893.

33 As part of his efforts to discredit alternating current (AC) power, Edison took to publicly electrocuting stray animals and—although in principle opposed to capital punishment—tasked his research team with helping to develop the electric chair, using alternating current to achieve its deadly end.

34 For more on Binder's design philosophy, see Joseph Binder, *Color in Advertising* (London and New York: Studio Publications, 1934). For an overview of his life and work, including an illustrated selection of additional posters in the work safety series, see *Joseph Binder: Wien–New York*, ed. Peter Noever (Vienna: Österreichisches Museum für angewandte Kunst, 2001).

35 It is worth noting that Austrian and German collection databases often use the word *aufklärung*—meaning education (colloquially), reconnaissance (in technical use), and, significantly, the Enlightenment—as a subject designation for posters in the series.

36 *Putting Progress through its Paces: The Story of the General Motors Proving Grounds*, 5th ed. (Detroit: General Motors Corporation, 1937), 15.

37 Ibid., 6.

38 Ibid., 29.

39 "Children Parade Against Death Ave.," *New York Times*, October 25, 1908.

40 "At Cornell Brains Are Dissected to Find the Mechanism of Thought," *Life*, July 24, 1939, 48–49.

PLATES

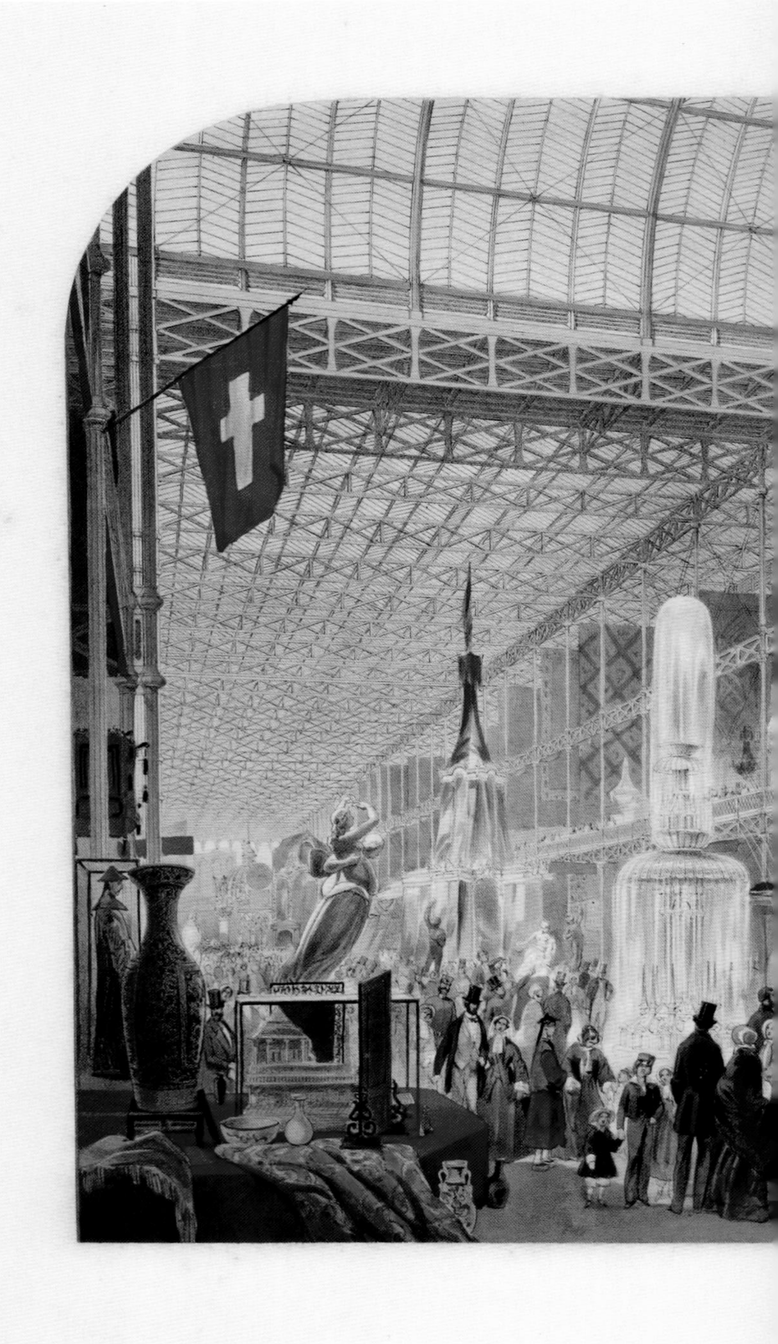

Pl. 1
"General View of the Interior," from *Recollections of the Great Exhibition*, 1851.
After John Absolon (1815–1895) and William Telbin (1813–1873)

PERSIA

Pl. 2

Poster, *Exposition Universelle et Internationale de Bruxelles 1910* (Universal and International Exhibition, Brussels 1910), 1909–10. Henri Bellery-Desfontaines (1867–1909), designer

Pl. 3
Vase, 1910. Gordon Mitchell Forsyth (1879–1952), designer and decorator

Pl. 4
Vase, 1910. Gordon Mitchell Forsyth (1879–1952), designer and decorator

Pl. 5
Sticker, *Helft der Glaspalast Künstlerhilfe* (Help the Glass Palace Artist Assistance), 1931. Ludwig Hohlwein (1874–1949), designer

Pl. 6
Painting, *Die Verschollenen* (The Missing), 1934. Edgar Ende (1901–1965)

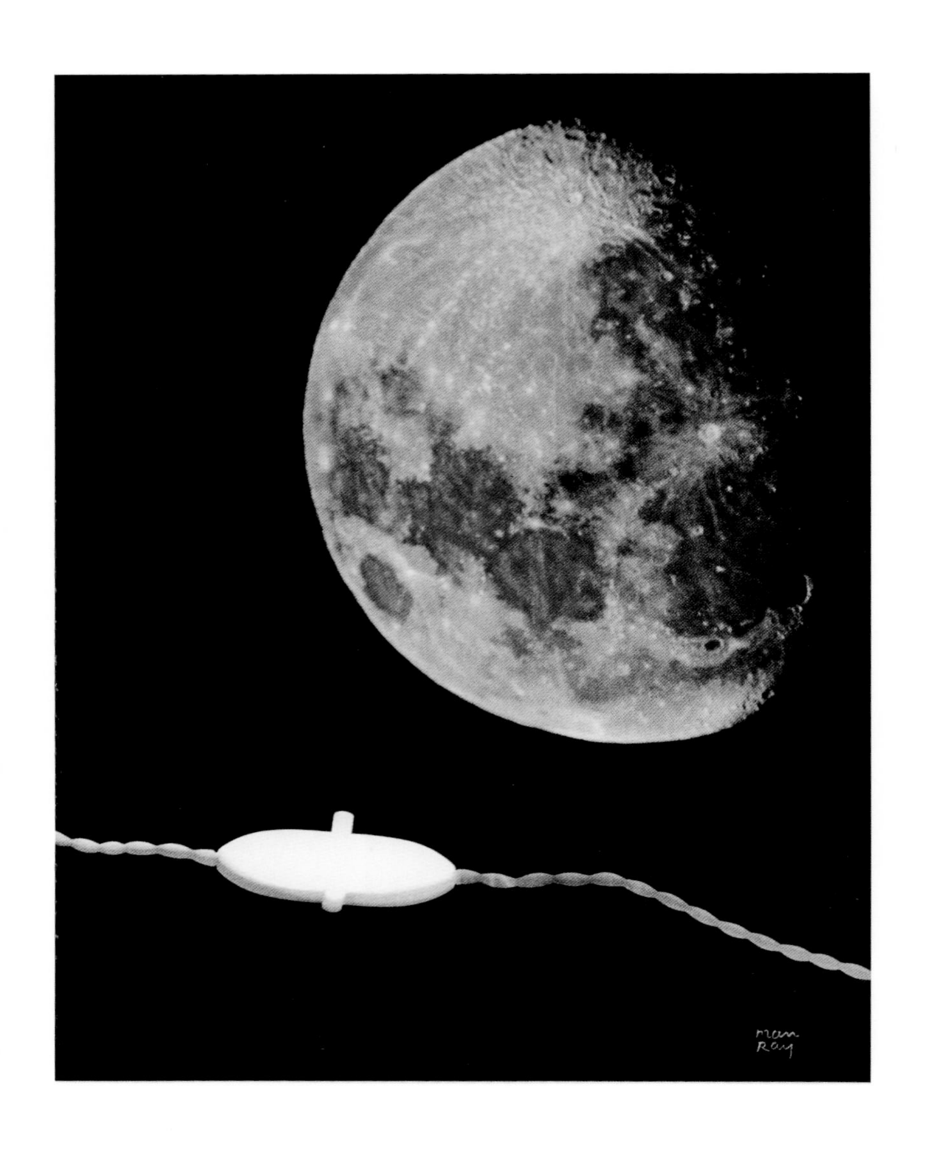

Pl. 7

Le monde (The World), 1931. Man Ray (1890–1976), photographer

Pl. 8
Électricité (Electricity), 1931. Man Ray (1890–1976), photographer

Pl. 9
La ville (The City), 1931. Man Ray (1890–1976), photographer

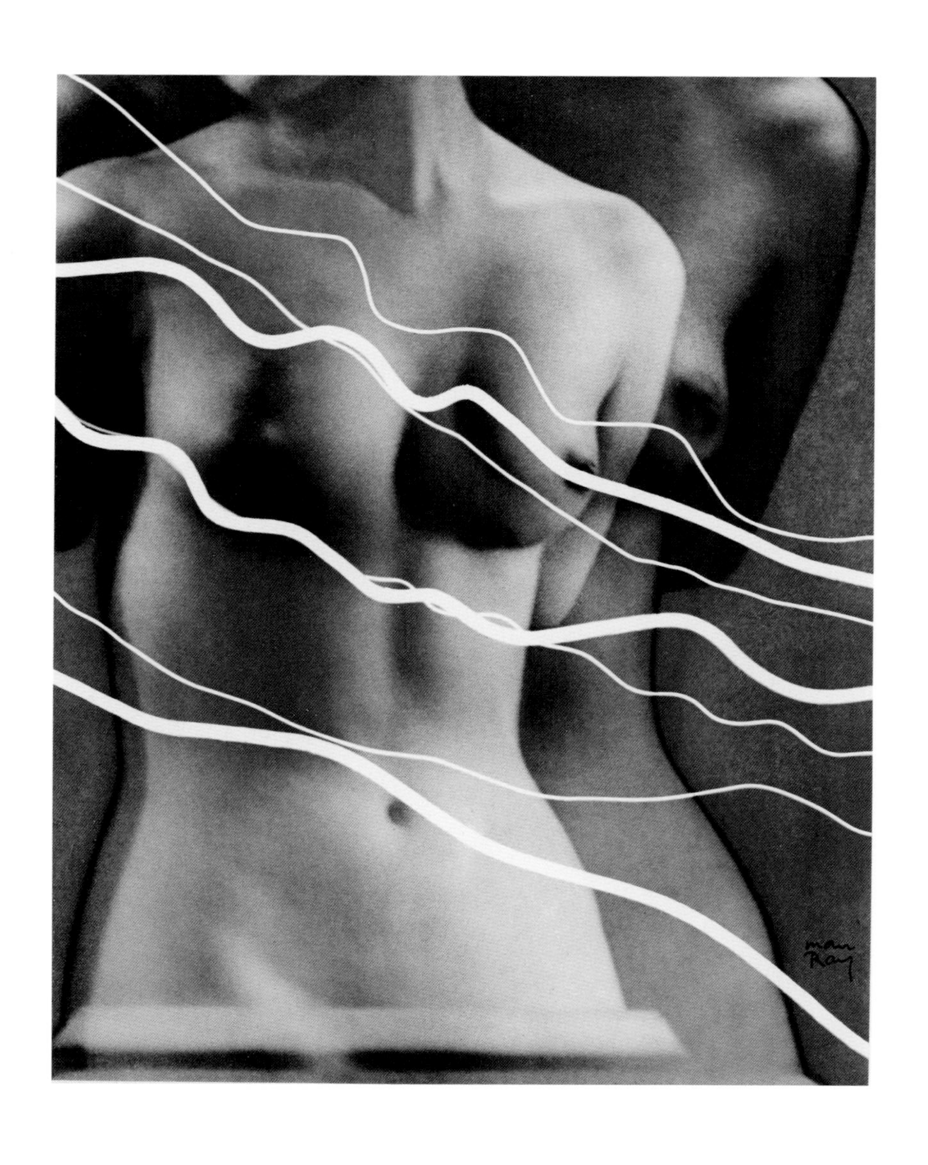

Pl. 10
Électricité (Electricity), 1931. Man Ray (1890–1976), photographer

Pl. 11

Poster, *Onoranze a Volta nel centenario della pila* (Honors to Volta on the Centenary of the Battery), 1898. Adolfo Hohenstein (1854–1928), designer

Pl. 12

Poster, *Tentoonstelling van Electriciteit in Huis en Ambacht* (Exhibition of Electricity in the Home and Handwork), 1907. Daan (Daniël) Hoeksema (1879–1935), designer

Pl. 13

Poster, *Enschede Zevenmijls Electriciteitstentoonstelling* (Enschede Seven Mile Electricity Exhibition), 1930. Job Denijs (1893–1970), designer

Pl. 14–15

Stickers, *Electrical and Industrial Exposition*, 1922, 1928

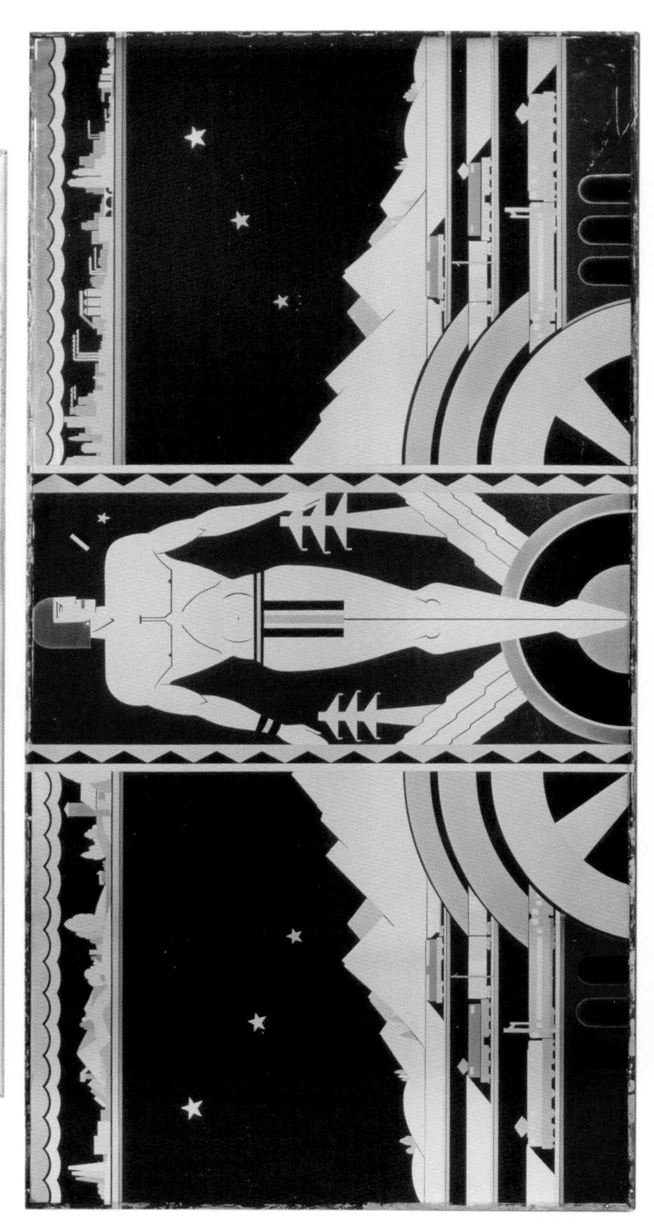

Pl. 16

Panels, *High Voltage Railway Electrification*, 1933. Center panel: Westinghouse Pavilion, 1933 Chicago A Century of Progress International Exposition; top and bottom panels: Westinghouse Electric and Manufacturing Company offices, Pittsburgh, c. 1935. Donald R. Dohner (1892–1943), designer

Pl. 17
Poster, *Op blanke draden loert gevaar* (Danger Lurks on Exposed Wires), 1939.
Eduard Strelitskie (1908–1995), designer

Pl. 18

Print, *Rückkoppler!! Schonet die Nerven eurer Mitmenschen!* (Feedback!! Spare the Nerves of Your Fellow Man!), 1927. Julius Klinger (1876–after 1942), designer

Pl. 19

Poster, *Gib acht sonst . .* (Be Careful or Else . . .), 1929–30. Joseph Binder (1898–1972), designer

Pl. 20
Poster, *Vioittunut käsilamppu on kalman kädenpuristus* (A Damaged Hand Lamp Is a Deadly Handshake), 1932. R. Saavzn, designer

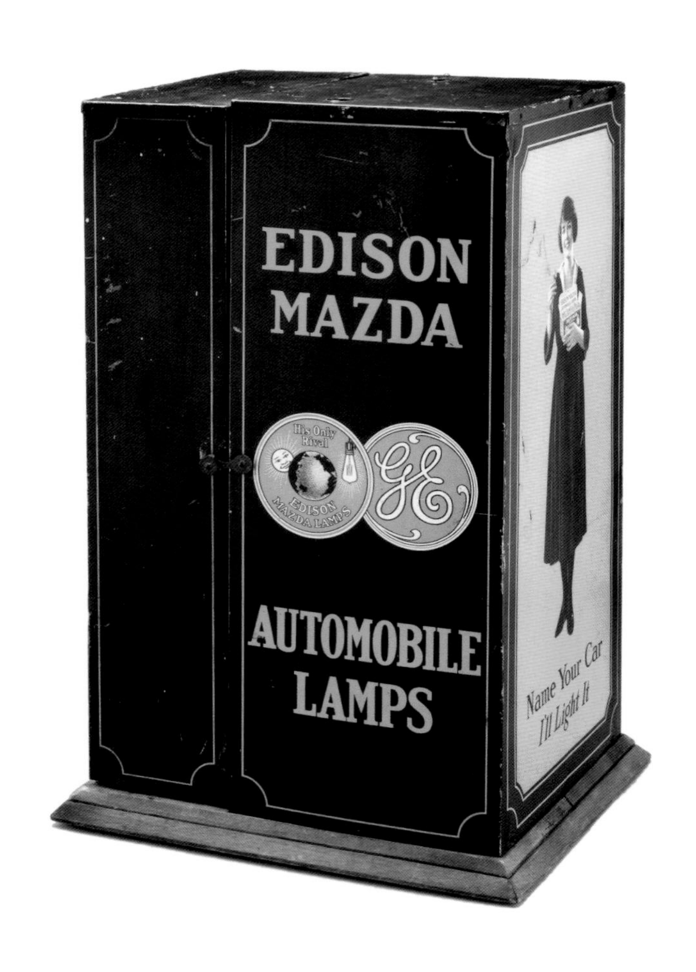

Pl. 21
Display case for Edison Mazda automobile lamps, c. 1923

Pl. 22
Poster, *Philips Duplo en Triplo lampen verblinden niet* (Philips Duplo and Triplo Lamps Do Not Blind), 1928. Mathieu (Nicolaas Cornelis) Clement (1905–1929), designer

Pl. 23–24

Posters, from the series *Teekenen en kleuren voor de lagere school* (Drawing and Colors for Primary School), Series B, 1929. A. Posma, K. A. Smit, and Wiebe Cornel, designers

Koorts

het geslaagd zijn. Het is werkelijk overdreven, dat de menschen zich voor die rijproef nog zoo geweldig kunnen opwinden. We hebben candidaten meegemaakt, die zaten te rillen en te beven en wien het angstzweet uitbrak. Nu is het waar, dat er voor sommigen heel wat van afhangt. Er zijn menschen, die lang werkloos geweest zijn en nu

169

Pl. 25

"Examen koorts" (Feverish Exam), from *100,000 Kilometer van wielen en wegen* (100,000 Kilometers of Wheels and Roads), c. 1940. Piet Marée (1903–1999), designer

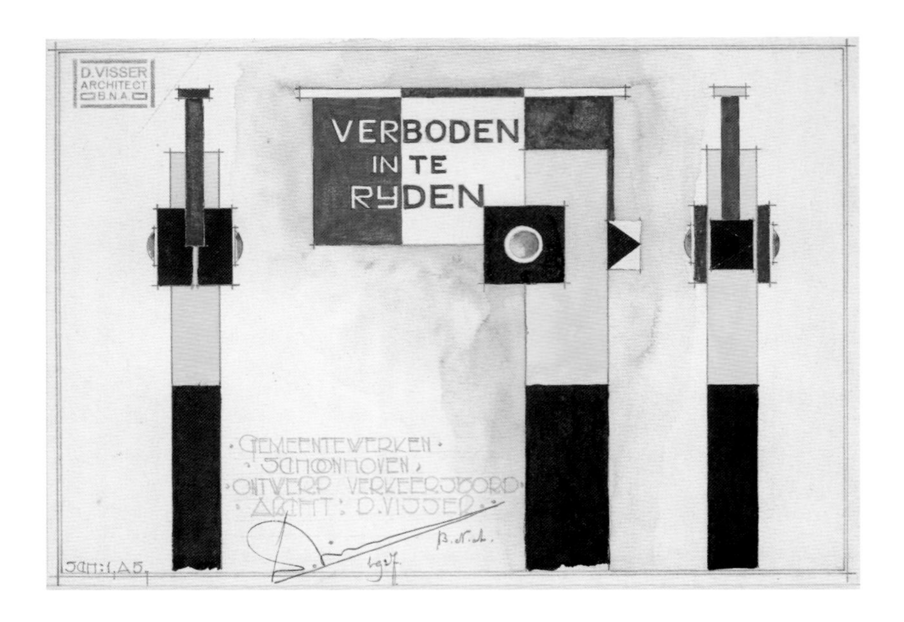

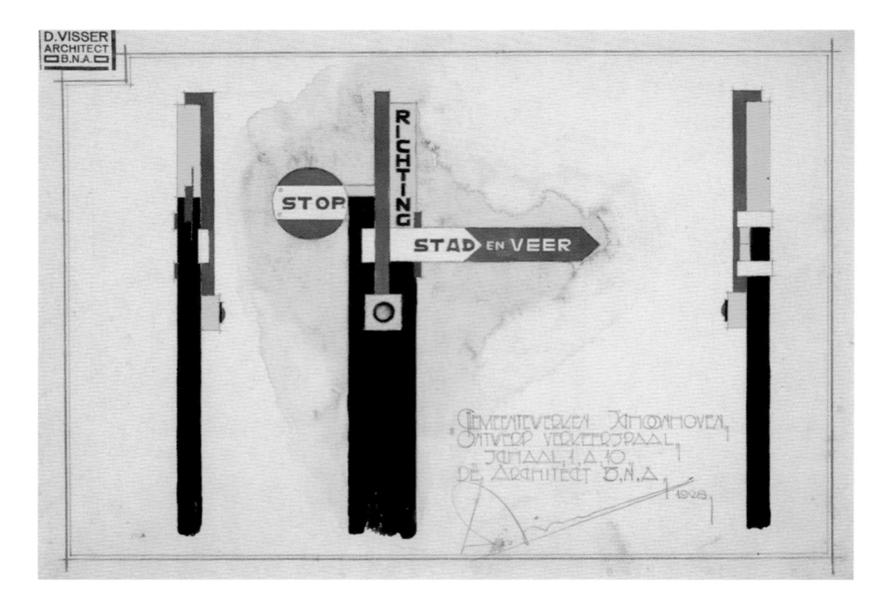

Pl. 26–27

Designs for traffic signals, for the Public Works Department, Schoonhoven, the Netherlands. Died Visser (1899–1977), designer

Verboden in te rijden (Do Not Enter), 1927

Stop. Richting. Stad en veer (Stop. Direction. City and Ferry), 1928

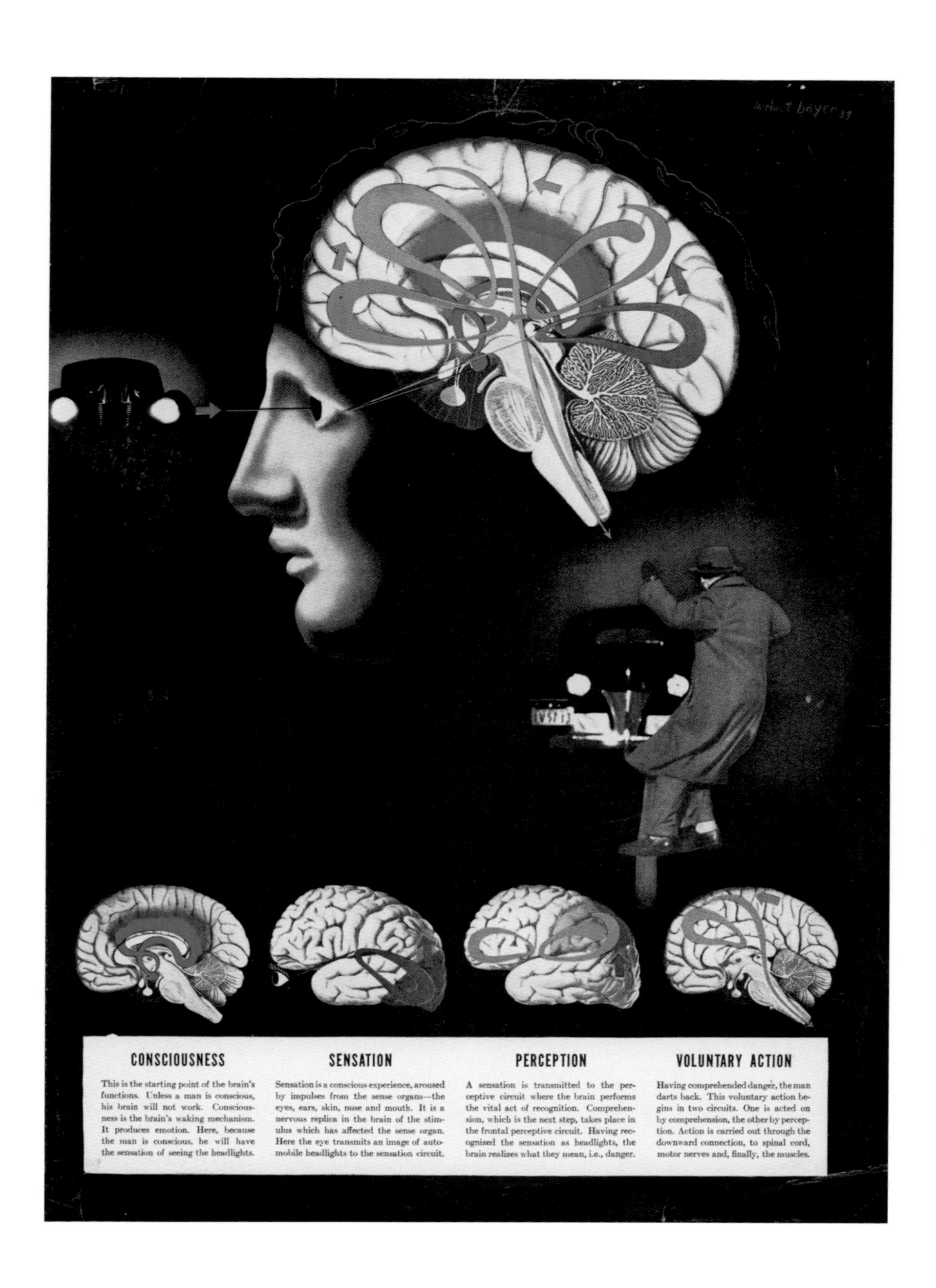

Pl. 28
Proof, "Consciousness, Sensation, Perception, Voluntary Action," for *Life*, July 24, 1939. Herbert Bayer (1900–1985), designer

Pl. 29

Poster, *Onveilige stempelpersen stanzen, e.d.!! 546 ongevallen per jaar* (Unsafe Stamp Presses and Punches, etc.!! 546 Accidents per Year), 1940. Jan (Johannes Frederik) Lavies (1902–2005), designer

Pl. 30
Poster, *Pas op die braam!* (Beware of the Burr!), 1940. Endre Lukács (1906–2001), designer

Pl. 31

Poster, *Gesleten kabel, kapotte handen* (Worn-Out Cable, Broken Hands), 1942. R. Wormer, designer

Pl. 32

Poster, *Dit kan een voet kosten* (This Can Cost You a Foot), 1940. Endre Lukács (1906–2001), designer

Pl. 33
Poster, *Slijp veilig. Één splinter kan u een oog kosten* (Grind Safely. One Splinter Can Cost You an Eye), 1942. Hans (Johannes Hendrikus) Bolleman (1923–1968), designer

Pl. 34

Poster, *Iedere verstandige lasser gebruikt een veiligheidsbril* (Every Sensible Welder Uses Safety Goggles), 1950 (designed 1940). Jacob Jansma (1893–1972), designer

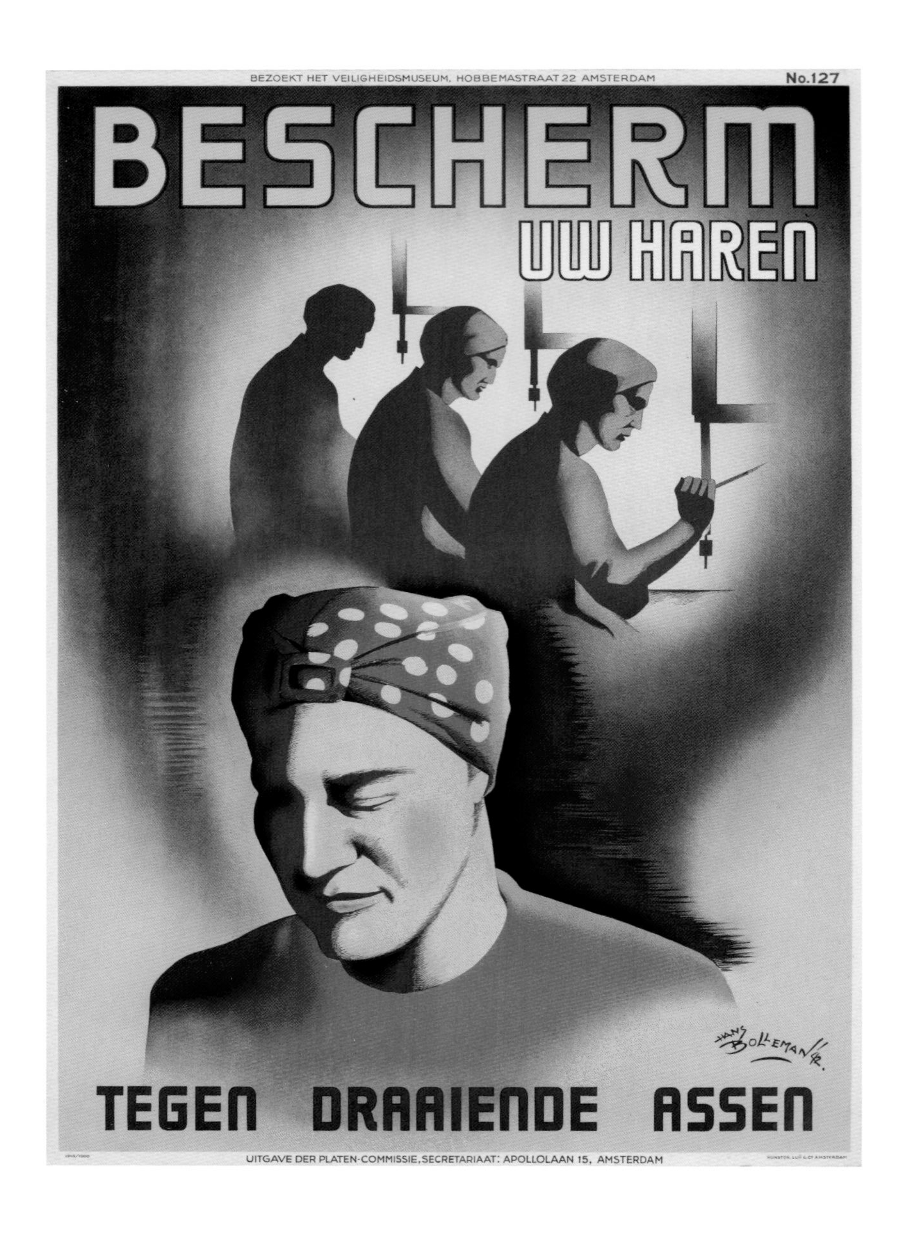

Pl. 35

Poster, *Bescherm uw haren tegen draaiende assen* (Protect Your Hair Against Revolving Spindles), 1943 (designed 1942). Hans (Johannes Hendrikus) Bolleman (1923–1968), designer

Pl. 36

Poster, *Werk veilig! Denk aan moeder* (Work Safely! Think of Mother), 1942.
Hans (Johannes Hendrikus) Bolleman (1923–1968), designer

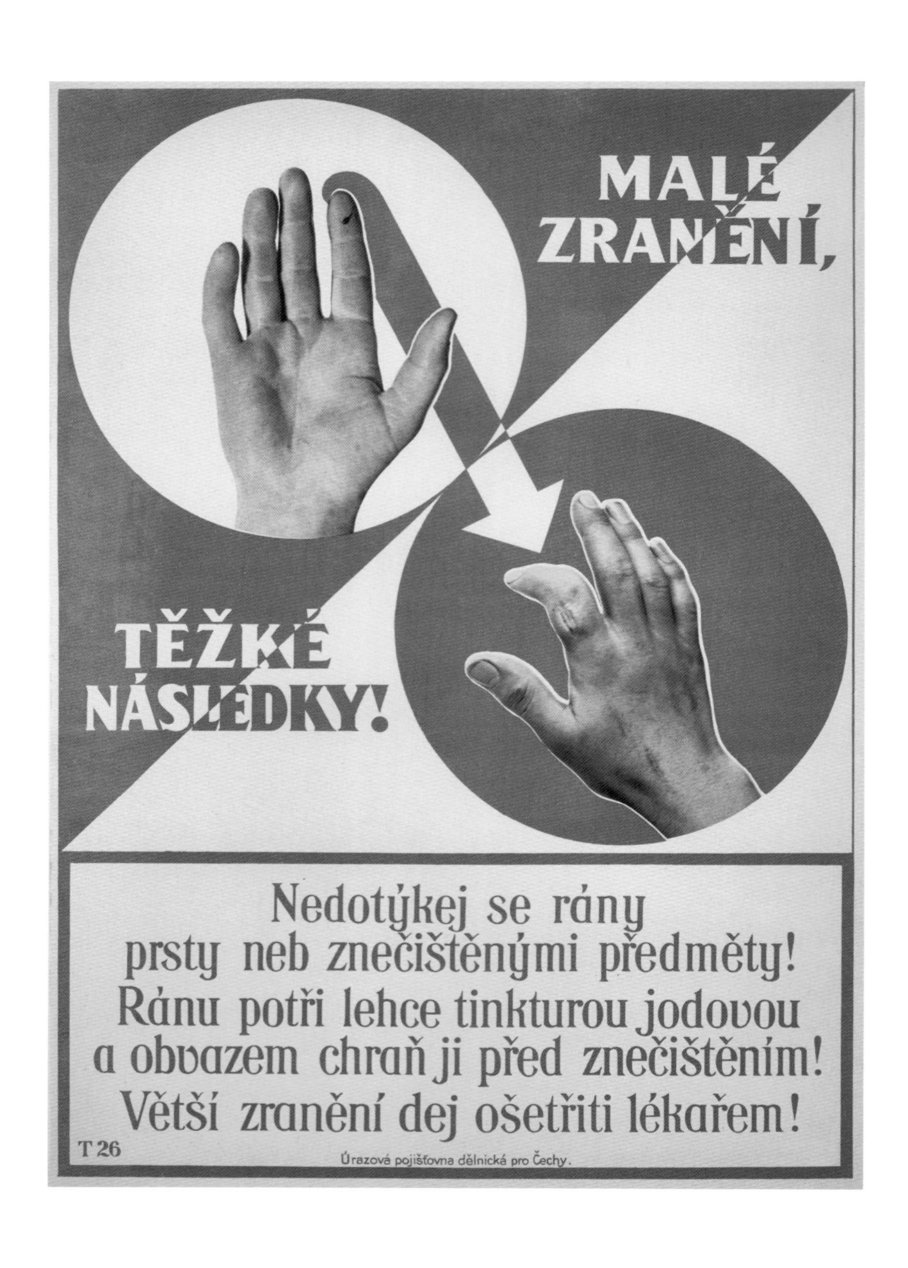

Pl. 37
Poster, *Malé zranění, těžké následky!* (Small Injuries, Serious Consequences!), c. 1934

Pl. 38

Poster, *Zastav stroj! Neodstraňuj odpadkov dokiaľ stroj beží!* (Stop the Engine! Do Not Remove Scraps while the Engine Is Running!), c. 1934

Pl. 39

Poster, *Zdvižená kukla před úrazem nechrání! Braní odříznutého kusu zpět jest vždy nebezpečné! Řiď se dle toho!* (A Raised Guard Does Not Protect Against Injury! Taking the Cutting Back Is Always Dangerous! Follow This Rule!), c. 1934

Pl. 40

Poster, *Neodstraňuj třísek rukou! I malé zranění je nebezpečné!* (Even a Small Injury Is Dangerous! Do Not Remove Splinters by Hand!), c. 1934

Pl. 41

Poster, *Zajištěnim točnice předejdeš úrazu!* (A Secured Turntable Prevents Accidents!), c. 1934

Pl. 42

Poster, *Odstraň nebo zatluč vyčnívající hřeby!* (Remove or Hammer Protruding Nails!), c. 1934

Pl. 43
Poster, *Nepořádek příčinou úrazu!* (Disorderliness Causes Accidents!), c. 1934

Pl. 44
Poster, *Pozor na vlak! Jde o tvůj život!* (Beware of the Train! It's Your Life on the Line!), c. 1934

A VLAK!

U ŽIVOT!

O ČECHY v PRAZE.

Pl. 45
Postcard, *Capelli sciolti, pericoli molti* (Loose Hair, Much Danger), 1938

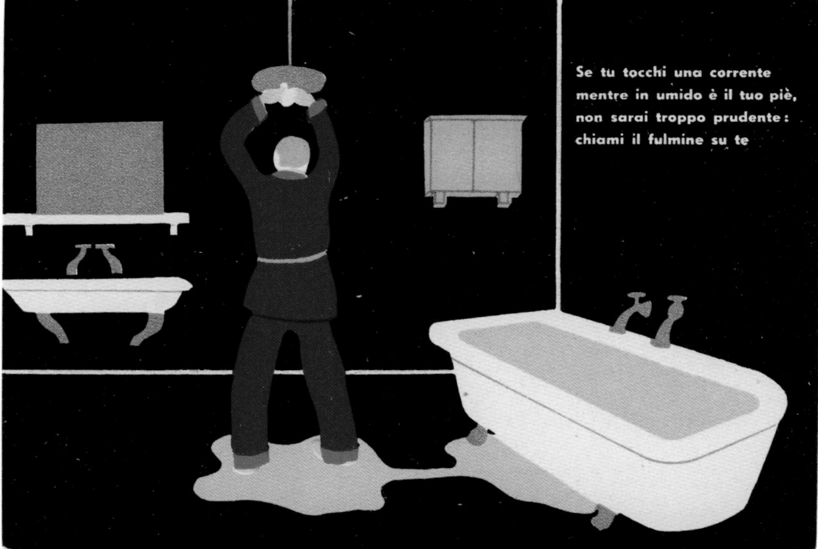

Pl. 46

Postcard, *Bada alla cinghia: guai se t'avvinghia!* (Pay Attention to the Belt: Woe If It Catches You!), 1938

Pl. 47

Postcard, *Se tu tocchi una corrente mentre in umido è il tuo piè, non sarai troppo prudente: chiami il fulmine su te* (If You Touch a Current While Your Foot Is Wet, You Would Not Have Been Too Prudent: You Call the Lightning Toward Yourself), 1938

Pl. 48

Postcard, *Olio sulla pista, ospedale in vista* (Oil on the Track, Hospital on the Horizon), 1938

Pl. 49

Postcard, *Getti un fiammifero dove va, va? È prevedibile quel che accadrà* (You Throw a Match, Wherever It Goes, It Goes. What Will Happen You Can Foresee), 1938

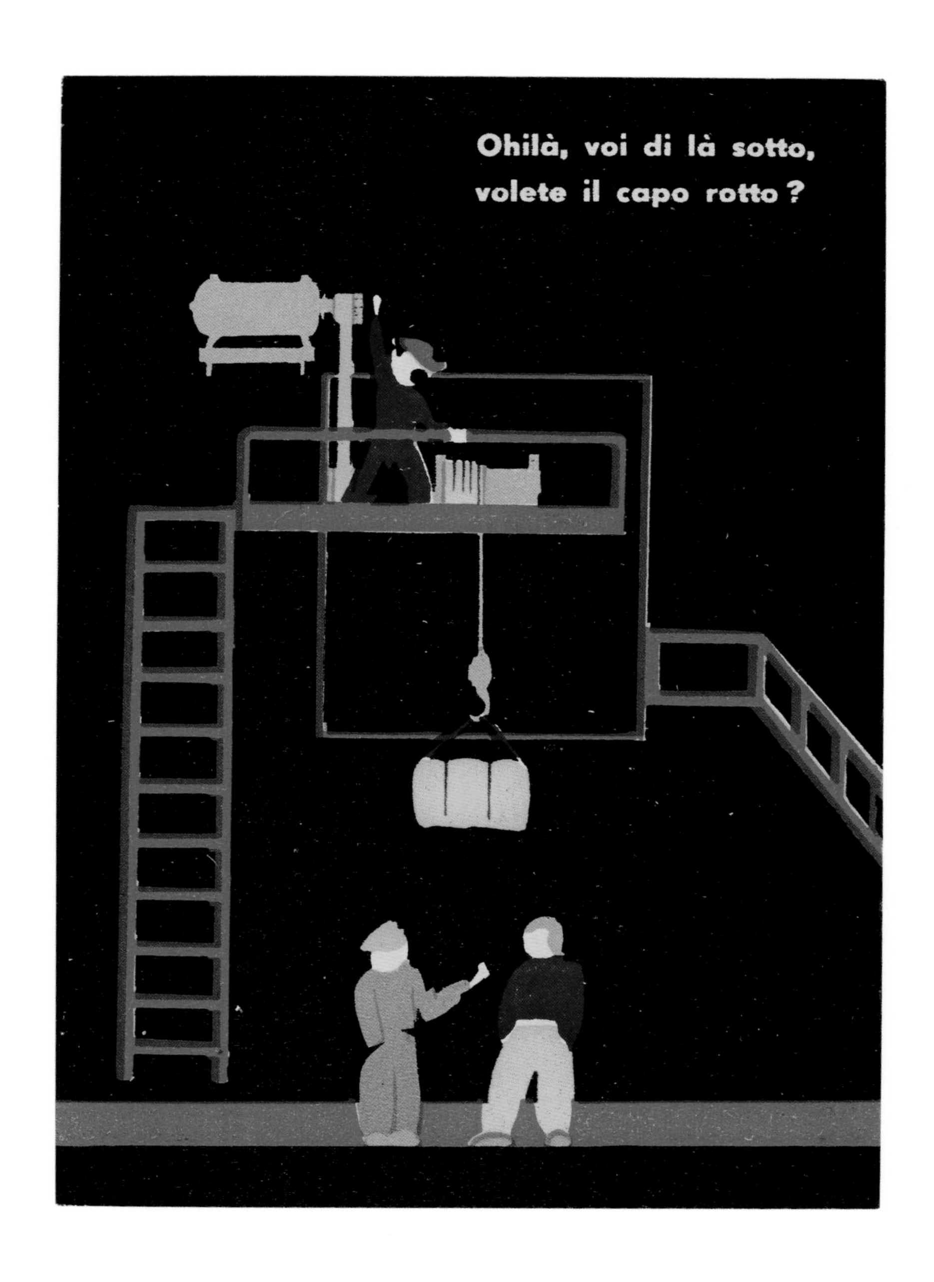

Pl. 50

Postcard, *Ohilà, voi di là sotto, volete il capo rotto?* (Hey, You Down There, Do You Want Your Head to Be Broken?), 1938

Pl. 51
Postcard, *Chi vuol salir così cade sovente, Precipitevolissimevolmente* (He Who Wants to Climb This Way, Often Falls Headlong), 1938

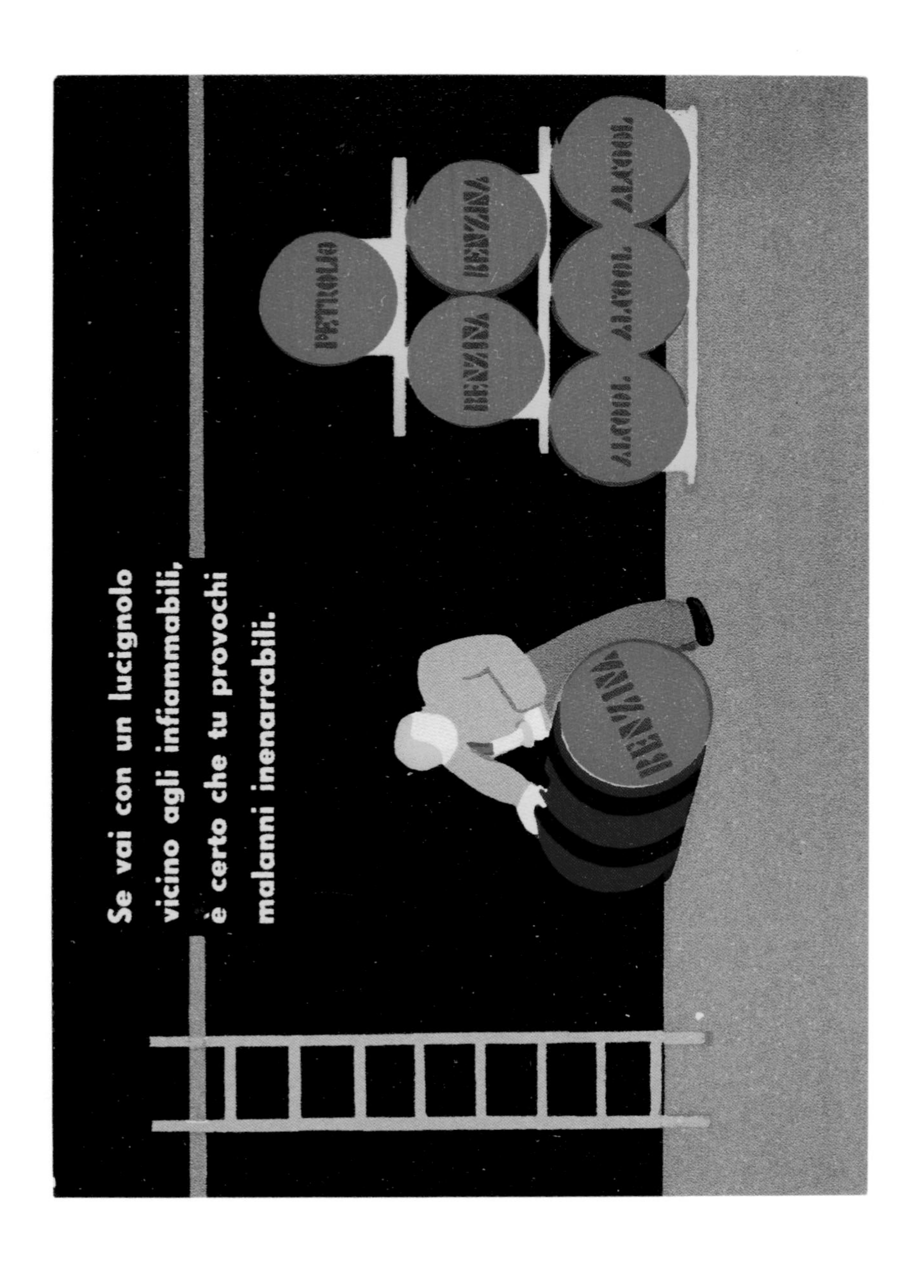

Pl. 52

Postcard, *Se vai con un lucignolo vicino agli infiammabili, è certo che tu provochi malanni inenarrabili* (If You Hold a Wick near Flammable Material, It Is Certain That You Will Provoke Unspeakable Misfortunes), 1938

Pl. 53
Postcard, *Abito svolazzante, Pericolo imminente* (Fluttering Dress, Imminent Danger), 1938

Pl. 54
Postcard, *Se non vedi dove vai, Il piè in fallo metterai!* (If You Do Not See Where You Are Going, You Will Put Your Foot Afoul!), 1938

Pl. 55
Postcard, *Colui che lascia intoppi sul passaggio, Non è davvero un uomo accorto e saggio* (He Who Leaves Obstacles in the Way, Is Not at All a Shrewd and Wise Man), 1938

P1. 56
Postcard, *Annoda saldi i carichi, altrimenti saran per te pericoli e . . . accidenti* (Tie the Loads Firmly, Otherwise It Will Be Dangers and Accidents For You), 1938

Pl. 57
Poster, (Halt! Do Not Go to the Front Immediately after Blasting), c. 1924

Pl. 58
Poster, (It Is Dangerous to Jump into a Moving Cage), c. 1924

Pl. 59
Poster, (Ensure Your Safety with Secure Timbering), c. 1924

Pl. 60
Poster, (This Leather Hat Brings Protection), c. 1924

Pl. 61
Photograph, *Nerve and Steel*, 1933–35. Wendell MacRae (1896–1980)

All works from The Wolfsonian–FIU, The Mitchell Wolfson, Jr. Collection, unless otherwise noted.

FIGURES

1–2
Postcards, *Bruxelles-Exposition. L'incendie des 14–15 Août 1910* (Brussels Exposition. The Fire of 14–15 August 1910), 1910
3 1/2 x 5 1/2 in. (9 x 14 cm) each
XB1992.1581.2, .3

Panorama de Bruxelles-Kermesse après les ravages du feu (Panorama of Brussels Fair after the Ravages of Fire)

Bruxelles-Kermesse. Vers la rue de l'Escalier (Brussels Fair. Toward the Rue de l'Escalier)

3
View of the Crystal Palace at Sydenham, page from *The Crystal Palace to Be Sold by Auction . . .*, 1911
Hudson & Kearns, Ltd., London, printer
XC1993.533

4
Stereograph, *The Immense Dynamos, Machinery Hall, Columbian Exposition,* 1893
B. W. (Benjamin West) Kilburn (American, 1827–1909), photographer
B. W. Kilburn, Littleton, New Hampshire, publisher
3 1/2 x 7 1/8 in. (9 x 18 cm)
XB1998.40.41

5
Stereograph, *Illumination and Great Search Light, Columbian Exposition*, 1894
B. W. (Benjamin West) Kilburn (American, 1827–1909), photographer
B. W. Kilburn, Littleton, New Hampshire, publisher
3 1/2 x 7 1/8 in. (9 x 18 cm)
XB1998.40.53

6–7
Portfolio, *Électricité* (Electricity), 1931
Man Ray (American, 1890–1976), photographer
Pierre Bost (French, 1901–1975), author
La Compagnie Parisienne de Distribution d'Électricité, Paris, publisher
Photogravures
15 x 12 in. (38.1 x 30.5 cm)
86.2.146

Salle à manger (Dining Room)
Lingerie

8
Book cover, *Die Berliner Elektrizitätswerke bis Ende 1896* (The Berlin Electric Power Stations by the End of 1896), 1897
Ludwig Sütterlin (German, 1865–1917), cover designer
Gustav Kemmann (German, 1858–1931), author
Julius Springer, Berlin, and R. Oldenbourg, Munich, publishers
12 1/4 x 9 7/8 in. (31 x 25 cm)
TD1989.113.10

9
Photograph, *Osram*, c. 1935
John Havinden (British, 1908–1987)
London
Gelatin silver print
11 1/2 x 14 1/2 in. (29.2 x 36.8 cm)
86.21.34

10
"Westinghouse Electric and Manufacturing Company," page from *Official Pictures of A Century of Progress Exposition*, 1933
Reuben H. Donnelley Corporation, Chicago, publisher
83.2.15

11
"Eyes that Cast Shadows," page 14 from *Putting Progress through Its Paces: The Story of the General Motors Proving Grounds*, 1937
Otto Linstead (American, 1887–1970), photographer
Department of Public Relations, General Motors Corporation, Detroit, publisher
8 1/4 x 5 1/2 in. (21 x 14 cm)
The Wolfsonian–FIU, Purchase, XC2011.05.1.3

12
Photograph, *Silhouette – N.Y.C. Elevated Motor West Side under Construction near 14th St.*, 1930
Gordon Coster (American, 1906–1988)
New York City
Gelatin silver print
15 1/2 x 18 7/8 in. (34.9 x 42.9 cm)
XX1990.2480

13
"Aus dem Federnwerk: Wickeln von Flugmotoren-Ventilfedern" (From the Spring Works: Winding of Aircraft Engine Valve Springs), page 39 from *50 Jahre Poldihütte: Entwicklung, gegenwärtiger Stand, Arbeitsverfahren, Erzeugnisse* (50 Years Poldihütte: Development, Present Status, Working Method, Products), 1939
Poldihütte, Prague, publisher
XC1992.720

14
Print, *Mid-air*, 1931
Louis Lozowick (American, b. Ukraine, 1892–1973)
New York City
Lithograph
15 7/8 x 11 7/8 in. (40.3 x 30.2 cm)
84.4.129

PLATES

1
"General View of the Interior," plate 2 from *Recollections of the Great Exhibition*, 1851
After John Absolon (British, 1815–1895) and William Telbin (British, 1813–1873)
Lloyd Brothers & Co., London, publisher
Day & Son, Ltd., London, printer
Hand-colored lithograph
18 1/8 x 24 in. (46 x 61 cm)
XB1990.2296

2
Poster, *Exposition Universelle et Internationale de Bruxelles 1910* (Universal and International Exhibition, Brussels 1910), 1909–10
Henri Bellery-Desfontaines (French, 1867–1909), designer
Eugéne Verneau, Paris, printer
Offset lithograph
44 1/2 x 34 in. (113 x 86.4 cm)
TD1991.172.4

3–4
Vase, 1910
Gordon Mitchell Forsyth (British, 1879–1952), designer and decorator
Pilkington's Tile & Pottery Company, Clifton Junction, England, maker
Luster glazed earthenware
21 1/2 x 16 in., diam. (54.6 x 40.6 cm, diam.)
TD1991.144.1

5
Sticker, *Helft der Glaspalast Künstlerhilfe* (Help the Glass Palace Artist Assistance), 1931
Ludwig Hohlwein (German, 1874–1949), designer
Glaspalast-Künstlerhilfe, Munich, publisher
2 3/4 x 1 5/8 in. (6.9 x 4.1 cm)
XB1991.979

6
Painting, *Die Verschollenen* (The Missing), 1934
Edgar Ende (German, 1901–1965)
Munich
Oil on canvas
51 1/2 x 79 1/4 in. (130.8 x 201.3 cm)
XX1989.187

7–10
Portfolio, *Électricité* (Electricity), 1931
Man Ray (American, 1890–1976), photographer
Pierre Bost (French, 1901–1975), author
La Compagnie Parisienne de Distribution d'Électricité, Paris, publisher
Photogravures
15 x 12 in. (38.1 x 30.5 cm)
86.2.146

Le monde (The World)
Électricité (Electricity)
La ville (The City)
Électricité (Electricity)

11
Poster, *Onoranze a Volta nel centenario della pila* (Honors to Volta on the Centenary of the Battery), 1898
Adolfo Hohenstein (German, b. Russia, 1854–1928), designer
Officine G. Ricordi & C., Milan, publisher
Chromolithograph
43 x 20 3/8 in. (109.2 x 51.8 cm)
86.4.25

12
Poster, *Tentoonstelling van Electriciteit in Huis en Ambacht* (Exhibition of Electricity in the Home and Handwork), 1907
Daan (Daniël) Hoeksema (Dutch, 1879–1935), designer
Kluppell & Ebeling, Arnhem, printer
Offset lithograph
38 1/4 x 25 3/8 in. (97.2 x 64.5 cm)
86.4.50

13
Poster, *Enschede Zevenmijls Electriciteitstentoonstelling* (Enschede Seven Mile Electricity Exhibition), 1930
Job Denijs (Dutch, 1893–1970), designer
Drukkerij Kotting, Amsterdam, printer
Offset lithograph
48 x 27 3/8 in. (121.9 x 69.5 cm)
XX1992.185

14
Sticker, *Electrical and Industrial Exposition*, 1922
2 3/4 x 1 3/4 in. (7 x 4.5 cm)
XB2005.10.1.7

15
Sticker, *Electrical and Industrial Exposition*, 1928
2 1/2 x 1 7/8 in. (6.5 x 4.8 cm)
XB2005.10.1.40

16
Panels, *High Voltage Railway Electrification,* 1933
Center panel: Westinghouse Pavilion, 1933 Chicago A Century of Progress International Exposition; top and bottom panels: Westinghouse Electric and Manufacturing Company offices, Pittsburgh, c. 1935
Donald R. Dohner (American, 1892–1943), designer
Westinghouse Electric and Manufacturing Company, Pittsburgh, manufacturer
Micarta and dyed aluminum, steel, wood
Center panel: 49 1/2 x 97 5/8 x 1 1/4 in. (125.7 x 248 x 3.2 cm); top and bottom panels: 6 x 78 1/4 x 1 3/8 in. (15.2 x 198.7 x 3.5 cm) each
83.6.11 a–c

17
Poster, *Op blanke draden loert gevaar* (Danger Lurks on Exposed Wires), 1939
Eduard Strelitskie (Dutch, 1908–1995), designer
Platen-Commissie, Rijksverzekeringsbank, Amsterdam, publisher
L. Van Leer & Co., Amsterdam, printer
Offset lithograph
26 7/8 x 19 3/8 in. (68.3 x 49.2 cm)
TD1990.123.9

18
Print, *Rückkoppler!! Schonet die Nerven eurer Mitmenschen!* (Feedback!! Spare the Nerves of Your Fellow Man!), 1927
Julius Klinger (Austrian, 1876–after 1942), designer
Offset lithograph
9 x 5 7/8 in. (23 x 15 cm)
XX1990.2613

19
Poster, *Gib acht sonst . .* (Be Careful or Else . . .), 1929–30
Joseph Binder (American, b. Austria, 1898–1972), designer
Österreichische Zentralstelle für Unfallverhütung, Vienna, publisher
Ferdinand Kehlborn, Vienna, printer
Offset lithograph
16 5/8 x 11 5/8 in. (42.2 x 29.5 cm)
87.4.8

20
Poster, *Vioittunut käsilamppu on kalman kädenpuristus* (A Damaged Hand Lamp Is a Deadly Handshake), 1932
R. Saavzn, designer
Suomen Tapaturmavakuutuslaitosten Liiton, Helsinki, publisher
Arvidsson Lito, Helsinki, printer
Offset lithograph
23 3/8 x 16 1/2 in. (59.4 x 41.9 cm)
87.4.5

21
Display case for Edison Mazda automobile lamps, c. 1923
General Electric, New York, manufacturer
Enameled tin, wood
24 5/8 x 16 x 13 7/8 in. (62.5 x 40.6 x 35.2 cm)
86.20.15

22
Poster, *Philips Duplo en Triplo lampen verblinden niet* (Philips Duplo and Triplo Lamps Do Not Blind), 1928
Mathieu (Nicolaas Cornelis) Clement (Dutch, 1905–1929), designer
Philips, Eindhoven, publisher
Offset lithograph
38 1/4 x 25 3/4 in. (97.2 x 65.4 cm)
87.4.27

23–24
Posters, from the series *Teekenen en kleuren voor de lagere school* (Drawing and Colors for Primary School), Series B, 1929
A. Posma, K. A. Smit, and Wiebe Cornel, designers
Ten Brink, Arnhem, publisher
Offset lithographs
23 1/8 x 17 1/4 in. (58.7 x 43.8 cm) each
The Wolfsonian–FIU, The Mitchell Wolfson, Jr. Collection of Decorative and Propaganda Arts, Promised Gift, WC2006.5.11.38, .36

25
"Examen koorts" (Feverish Exam), page 169 from *100,000 Kilometer van wielen en wegen* (100,000 Kilometers of Wheels and Roads), c. 1940
Piet Marée (Dutch, 1903–1999), designer
H. J. Peppink (Dutch), author
Zuid-Hollandsche, The Hague, publisher
XX1990.505

26
Design for a traffic signal, *Verboden in te rijden* (Do Not Enter), 1927
For the Public Works Department, Schoonhoven, the Netherlands
Died Visser (Dutch, 1899–1977), designer
Watercolor, gouache, and graphite on paper
9 3/4 x 13 1/4 in. (24.8 x 33.7 cm)
XX1990.4240

27
Design for a traffic signal, *Stop. Richting. Stad en veer* (Stop. Direction. City and Ferry), 1928
For the Public Works Department, Schoonhoven, the Netherlands
Died Visser (Dutch, 1899–1977), designer
Watercolor, gouache, and graphite on paper
9 1/2 x 13 3/8 in. (24.1 x 34 cm)
XX1990.2071

28
Proof, "Consciousness, Sensation, Perception, Voluntary Action," for *Life*, July 24, 1939
Herbert Bayer (American, b. Austria, 1900–1985), designer
Offset lithograph
14 x 10½ in. (35.5 x 26.7 cm)
XB1999.210.120

29
Poster, *Onveilige stempelpersen stanzen, e.d.!! 546 ongevallen per jaar* (Unsafe Stamp Presses and Punches, etc.!! 546 Accidents per Year), 1940
Jan (Johannes Frederik) Lavies (Dutch, 1902–2005), designer
Platen-Commissie, Rijksverzekeringsbank, Amsterdam, publisher
L. Van Leer & Co., Amsterdam, printer
Offset lithograph
27 x 19⅜ in. (68.6 x 49.2 cm)
TD1990.123.8

30
Poster, *Pas op die braam!* (Beware of the Burr!), 1940
Endre Lukács (Hungarian, 1906–2001), designer
Platen-Commissie, Rijksverzekeringsbank, Amsterdam, publisher
Luii & Co., Amsterdam, printer
Offset lithograph
26¾ x 19¼ in. (67.9 x 48.9 cm)
TD1990.123.4

31
Poster, *Gesleten kabel, kapotte handen* (Worn-Out Cable, Broken Hands), 1942
R. Wormer, designer
Platen-Commissie, Rijksverzekeringsbank, Amsterdam, publisher
Luii & Co., Amsterdam, printer
Offset lithograph
27 x 19½ in. (68.6 x 49.5 cm)
TD1990.123.10

32
Poster, *Dit kan een voet kosten* (This Can Cost You a Foot), 1940
Endre Lukács (Hungarian, 1906–2001), designer
Platen-Commissie, Rijksverzekeringsbank, Amsterdam, publisher
Luii & Co., Amsterdam, printer
Offset lithograph
26¾ x 19¼ in. (67.9 x 48.9 cm)
TD1989.318.2

33
Poster, *Slijp veilig. Één splinter kan u een oog kosten* (Grind Safely. One Splinter Can Cost You an Eye), 1942
Hans (Johannes Hendrikus) Bolleman (Dutch, 1923–1968), designer
Platen-Commissie, Rijksverzekeringsbank, Amsterdam, publisher
Luii & Co., Amsterdam, printer
Offset lithograph
26¾ x 19¼ in. (67.9 x 48.9 cm)
TD1989.318.1

34
Poster, *Iedere verstandige lasser gebruikt een veiligheidsbril* (Every Sensible Welder Uses Safety Goggles), 1950 (designed 1940)
Jacob Jansma (Dutch, 1893–1972), designer
Platen-Commissie, Rijksverzekeringsbank, Amsterdam, publisher
Luii & Co., Amsterdam, printer
Offset lithograph
27 x 19½ in. (68.6 x 49.5 cm)
TD1990.123.2

35
Poster, *Bescherm uw haren tegen draaiende assen* (Protect Your Hair Against Revolving Spindles), 1943 (designed 1942)
Hans (Johannes Hendrikus) Bolleman (Dutch, 1923–1968), designer
Platen-Commissie, Rijksverzekeringsbank, Amsterdam, publisher
Luii & Co., Amsterdam, printer
Offset lithograph
26¾ x 19⅜ in. (67.9 x 49.2 cm)
TD1990.123.6

36
Poster, *Werk veilig! Denk aan moeder* (Work Safely! Think of Mother), 1942
Hans (Johannes Hendrikus) Bolleman (Dutch, 1923–1968), designer
Platen-Commissie, Rijksverzekeringsbank, Amsterdam, publisher
Luii & Co., Amsterdam, printer
Offset lithograph
26⅞ x 19½ in. (68.3 x 49.5 cm)
TD1990.123.7

37
Poster, *Malé zranění, těžké následky!* (Small Injuries, Serious Consequences!), c. 1934
Úrazová pojišťovna dělnická pro Čechy, Prague, publisher
Offset lithograph
23⅜ x 16⅝ in. (59.4 x 42.2 cm)
The Wolfsonian–FIU, The Mitchell Wolfson, Jr. Collection of Decorative and Propaganda Arts, Promised Gift, WC2001.6.14.8

38
Poster, *Zastav stroj! Neodstraňuj odpadkov dokiaľ stroj beží!* (Stop the Engine! Do Not Remove Scraps while the Engine Is Running!), c. 1934
Spolok pre zábranu úrazov, Bratislava, publisher; originally issued by Úrazová pojišťovna dělnická pro Čechy, Prague
Offset lithograph
23 3/8 x 16 5/8 in. (59.4 x 42.2 cm)
The Wolfsonian–FIU, The Mitchell Wolfson, Jr. Collection of Decorative and Propaganda Arts, Promised Gift, WC2001.6.14.6

39
Poster, *Zdvižená kukla před úrazem nechrání! Braní odříznutého kusu zpět jest vždy nebezpečné! Řiď se dle toho!* (A Raised Guard Does Not Protect Against Injury! Taking the Cutting Back Is Always Dangerous! Follow This Rule!), c. 1934
Úrazová pojišťovna dělnická pro Čechy, Prague, publisher
Offset lithograph
23 3/8 x 16 5/8 in. (59.4 x 42.2 cm)
The Wolfsonian–FIU, The Mitchell Wolfson, Jr. Collection of Decorative and Propaganda Arts, Promised Gift, WC2001.6.14.1

40
Poster, *Neodstraňuj třísek rukou! I malé zranění je nebezpečné!* (Even a Small Injury is Dangerous! Do Not Remove Splinters by Hand!), c. 1934
Úrazová pojišťovna dělnická pro Čechy, Prague, publisher
V. Neubert, Prague, printer
Offset lithograph
16 5/8 x 23 3/8 in. (42.2 x 59.4 cm)
The Wolfsonian–FIU, The Mitchell Wolfson, Jr. Collection of Decorative and Propaganda Arts, Promised Gift, WC2001.6.14.11

41
Poster, *Zajištěním točnice předejdeš úrazu!* (A Secured Turntable Prevents Accidents!), c. 1934
Úrazová pojišťovna dělnická pro Čechy, Prague, publisher
Offset lithograph
16 5/8 x 23 3/8 in. (42.2 x 59.4 cm)
The Wolfsonian–FIU, The Mitchell Wolfson, Jr. Collection of Decorative and Propaganda Arts, Promised Gift, WC2001.6.14.3

42
Poster, *Odstraň nebo zatluč vyčnívající hřeby!* (Remove or Hammer Protruding Nails!), c. 1934
Úrazová pojišťovna dělnická pro Čechy, Prague, publisher
Offset lithograph
23 3/8 x 16 5/8 in. (59.4 x 42.2 cm)
The Wolfsonian–FIU, The Mitchell Wolfson, Jr. Collection of Decorative and Propaganda Arts, Promised Gift, WC2001.6.14.4

43
Poster, *Nepořádek příčinou úrazu!* (Disorderliness Causes Accidents!), c. 1934
Úrazová pojišťovna dělnická pro Čechy, Prague, publisher
Offset lithograph
23 3/8 x 16 5/8 in. (59.4 x 42.2 cm)
The Wolfsonian–FIU, The Mitchell Wolfson, Jr. Collection of Decorative and Propaganda Arts, Promised Gift, WC2001.6.14.5

44
Poster, *Pozor na vlak! Jde o tvůj život!* (Beware of the Train! It's Your Life on the Line!), c. 1934
Úrazová pojišťovna dělnická pro Čechy, Prague, publisher
M. Schulz, Prague, printer
Offset lithograph
16 5/8 x 23 3/8 in. (42.2 x 59.4 cm)
The Wolfsonian–FIU, The Mitchell Wolfson, Jr. Collection of Decorative and Propaganda Arts, Promised Gift, WC2001.6.14.13

45
Postcard, *Capelli sciolti, pericoli molti* (Loose Hair, Much Danger), 1938
Ente Nazionale di Propaganda per la Prevenzione degli Infortuni, publisher
Alterocca, Terni, Italy, printer
5 7/8 x 4 3/8 in. (15 x 11 cm)
86.19.800.6

46
Postcard, *Bada alla cinghia: guai se t'avvinghia!* (Pay Attention to the Belt: Woe If It Catches You!), 1938
Ente Nazionale di Propaganda per la Prevenzione degli Infortuni, publisher
Alterocca, Terni, Italy, printer
4 3/8 x 5 7/8 in. (11 x 15 cm)
86.19.800.5

47
Postcard, *Se tu tocchi una corrente mentre in umido è il tuo piè, non sarai troppo prudente: chiami il fulmine su te* (If You Touch a Current While Your Foot Is Wet, You Would Not Have Been Too Prudent: You Call the Lightning Toward Yourself), 1938
Ente Nazionale di Propaganda per la Prevenzione degli Infortuni, publisher
Alterocca, Terni, Italy, printer
4 3/8 x 5 7/8 in. (11 x 15 cm)
86.19.800.4

48
Postcard, *Olio sulla pista, ospedale in vista* (Oil on the Track, Hospital on the Horizon), 1938
Ente Nazionale di Propaganda per la Prevenzione degli Infortuni, publisher
Alterocca, Terni, Italy, printer
4 3/8 x 5 7/8 in. (11 x 15 cm)
86.19.800.3

49
Postcard, *Getti un fiammifero dove va, va? È prevedibile quel che accadrà* (You Throw a Match, Wherever It Goes, It Goes. What Will Happen You Can Foresee), 1938
Ente Nazionale di Propaganda per la Prevenzione degli Infortuni, publisher
Alterocca, Terni, Italy, printer
4 3/8 x 5 7/8 in. (11 x 15 cm)
86.19.800.2

50
Postcard, *Ohilà, voi di là sotto, volete il capo rotto?* (Hey, You Down There, Do You Want Your Head to Be Broken?), 1938
Ente Nazionale di Propaganda per la Prevenzione degli Infortuni, publisher
Alterocca, Terni, Italy, printer
5 7/8 x 4 3/8 in. (15 x 11 cm)
86.19.800.8

51
Postcard, *Chi vuol salir così cade sovente, Precipitevolissimevolmente* (He Who Wants to Climb This Way, Often Falls Headlong), 1938
Ente Nazionale di Propaganda per la Prevenzione degli Infortuni, publisher
Alterocca, Terni, Italy, printer
5 7/8 x 4 3/8 in. (15 x 11 cm)
86.19.800.7

52
Postcard, *Se vai con un lucignolo vicino agli infiammabili, è certo che tu provochi malanni inenarrabili* (If You Hold a Wick near Flammable Material, It Is Certain That You Will Provoke Unspeakable Misfortunes), 1938
Ente Nazionale di Propaganda per la Prevenzione degli Infortuni, publisher
Alterocca, Terni, Italy, printer
4 3/8 x 5 7/8 in. (11 X 15 cm)
86.19.800.1

53
Postcard, *Abito svolazzante, Pericolo imminente* (Fluttering Dress, Imminent Danger), 1938
Ente Nazionale di Propaganda per la Prevenzione degli Infortuni, publisher
Alterocca, Terni, Italy, printer
5 7/8 x 4 3/8 in. (15 x 11 cm)
86.19.800.9

54
Postcard, *Se non vedi dove vai, Il piè in fallo metterai!* (If You Do Not See Where You Are Going, You Will Put Your Foot Afoul!), 1938
Ente Nazionale di Propaganda per la Prevenzione degli Infortuni, publisher
Alterocca, Terni, Italy, printer
5 7/8 x 4 3/8 in. (15 x 11 cm)
86.19.800.10

55
Postcard, *Colui che lascia intoppi sul passaggio, Non è davvero un uomo accorto e saggio* (He Who Leaves Obstacles in the Way, Is Not at All a Shrewd and Wise Man), 1938
Ente Nazionale di Propaganda per la Prevenzione degli Infortuni, publisher
Alterocca, Terni, Italy, printer
5 7/8 x 4 3/8 in. (15 x 11 cm)
86.19.800.11

56
Postcard, *Annoda saldi i carichi, altrimenti saran per te pericoli e . . . accidenti* (Tie the Loads Firmly, Otherwise It Will Be Dangers and Accidents For You), 1938
Ente Nazionale di Propaganda per la Prevenzione degli Infortuni, publisher
Alterocca, Terni, Italy, printer
5 7/8 x 4 3/8 in. (15 x 11 cm)
86.19.800.12

57
Poster, (Halt! Do Not Go to the Front Immediately after Blasting), c. 1924
CSM, Belgium, publisher
Serigraph
22 5/8 x 28 3/4 in. (57.4 x 73 cm)
TD1988.170.55

58
Poster, (It Is Dangerous to Jump into a Moving Cage), c. 1924
CSM, Belgium, publisher
Serigraph
28 3/4 x 22 5/8 in. (73 x 57.4 cm)
TD1988.170.58

59
Poster, (Ensure Your Safety with Secure Timbering), c. 1924
CSM, Belgium, publisher
Serigraph
28 3/4 x 22 5/8 in. (73 x 57.4 cm)
TD1988.170.59

60
Poster, (This Leather Hat Brings Protection), c. 1924
CSM, Belgium, publisher
Serigraph
22 5/8 x 28 3/4 in. (57.4 x 73 cm)
TD1988.170.66

61
Photograph, *Nerve and Steel*, 1933–35
Wendell MacRae (American, 1896–1980)
New York City
Gelatin silver print
13 1/4 x 10 3/8 in. (33.7 x 26.4 cm)
84.21.10